Control and Estimation in Distributed Parameter Systems

Frontiers in Applied Mathematics

Frontiers in Applied Mathematics is a series that presents new mathematical or computational approaches to significant scientific problems. Beginning with Volume 4, the series reflects a change in both philosophy and format. Each volume focuses on a broad application of general interest to applied mathematicians as well as engineers and other scientists.

This unique series will advance the development of applied mathematics through the rapid publication of short, inexpensive books that lie on the cutting edge of research.

Frontiers in Applied Mathematics

Vol. 1	Ewing, Richard E., The Mathematics of Reservoir Simulation
Vol. 2	Buckmaster, John D., The Mathematics of Combustion
Vol. 3	McCormick, Stephen F., Multigrid Methods
Vol. 4	Coleman, Thomas F. and Van Loan, Charles, Handbook for Matrix Computations
Vol. 5	Grossman, Robert, Symbolic Computation: *Applications to Scientific Computing*
Vol. 6	McCormick, Stephen F., Multilevel Adaptive Methods for Partial Differential Equations
Vol. 7	Bank, R. E., PLTMG: A Software Package for Solving Elliptic Partial Differential Equations. *Users' Guide 6.0*
Vol. 8	Castillo, José E., Mathematical Aspects of Numerical Grid Generation
Vol. 9	Van Huffel, Sabine and Vandewalle, Joos, The Total Least Squares Problem: *Computational Aspects and Analysis*
Vol. 10	Van Loan, Charles, Computational Frameworks for the Fast Fourier Transform
Vol. 11	Banks, H.T. , Control and Estimation in Distributed Parameter Systems

Control and Estimation in Distributed Parameter Systems

Edited by H. T. Banks
North Carolina State University

Society for Industrial and Applied Mathematics
Philadelphia 1992

Library of Congress Cataloging-in-Publication Data

Control and estimation in distributed parameter systems / edited by
H. T. Banks.
p. cm. — (Frontiers in applied mathematics ; 11)
Includes bibliographical references and index.
ISBN 0-89871-297-1
1. Control theory. 2. Estimation theory. 3. Distributed
parameter systems. I. Banks, H. Thomas. II. Series.
QA402.3.C624 1992
629.8'312—dc20 92-27712

Contents

Contributors

A. Adamian, Mechanical, Aerospace and Nuclear Engineering, University of California, Los Angeles, California 90024

Ruth F. Curtain, Mathematics Institute, University of Groningen, Groningen, The Netherlands

M. C. Delfour, Centre de recherches mathématique et Département de mathématiques et statistique, Université de Montréal, Montréal, Québec H3C 3J7, Canada

J. S. Gibson, Mechanical, Aerospace and Nuclear Engineering, University of California, Los Angeles, California 90024

J. L. Lions, Mathématique des systèmes et de leur control, Collège de France, Paris (5), France

M. P. Polis, Electrical and Computer Engineering Department, Wayne State University, Detroit, Michigan 48202

David L. Russell, Department of Mathematics, Virginia Polytechnic Institute and State University, Blacksburg, Virginia 24061

Preface

The underlying idea for this Frontiers volume entailed convincing distinguished contributors to an increasingly important research area to write expository contributions on a broad range of topics fundamental to past, present, and future intellectual endeavors in the area of control of distributed parameter systems. Even though some of the resulting articles were completed more than four years ago, they are amazingly current with regard to importance and scope. Research in control and estimation of distributed parameter systems now encompasses a wide range of applications including both fundamental science and emerging technologies. The latter include smart materials (piezoceramics, shape memory alloys, magnetostrictives, electrorheological fluids) fabrication and testing, design of high-pressure chemical vapor deposition (CVD) reactors for production of microelectronic surfaces (e.g., semiconductors), while the former include groundwater contamination cleanup and other environmental modeling questions, climatology, flow control and fluid-structure interactions as well as more traditional topics in biology, mechanics, and acoustics. Fundamental research in mathematical topics is flourishing as is evidenced by the increasingly broad range of topics discussed at research gatherings of the mathematical control theory community (see, for example, [3],[4]) as well as the increasing numbers of high quality research papers in traditional journals such as the *SIAM J. Control and Optimization* and the *IEEE Transactions on Automatic Control* in addition to newer journals such as Mita Press's *Control-Theory and Advanced Technology* and Birkhäuser's *J. Mathematical Systems, Estimation and Control.* Even though the publication of this Frontiers volume has been delayed, the contributions are of a nature to provide substantial stimulus to both young researchers and experienced investigators in control theory.

In the first article, Lions considers pointwise control of distributed parameter systems and discusses a number of fundamental concepts including regularity, exact controllability using the by now well-known HUM (Hilbert Uniqueness Method) techniques, and optimality systems for both parabolic and hyperbolic systems. Pointwise actuator/sensor systems lead to unbounded input/unbounded output control and estimation problems which have become increasingly important in recent years. As Professor Lions points out, such systems are often only an idealization (in a limiting sense) of physical models for realizable actuator and sensors. However, in a growing list of important applications (in acoustics, flow control, flexible structures) where emerging technologies involving smart material actuators/sensors are used as control/observation devices, Dirac functions and their derivatives are in fact the most appropriate mathematical model for the physical system.

Both bounded and unbounded sensors and actuators play a role in the discussion of stabilization issues for large flexible structures in the second article by Delfour and Polis. They focus on hyperbolic systems and survey a number of open questions related to mathematical modeling in comparing sensors and actuators acting at a point or on the boundary versus distributed sensors and controls. Robustness of stabilization schemes (also a subject of Curtain's contribution in this volume) including the destabilizing effect of delays in boundary feedback are discussed.

A particular stabilization scheme, the linear quadratic regulator (LQR) feedback control design, for a distributed model of a flexible beam (with free end concentrated mass) attached to a rotating rigid hub is the focus of the paper by Gibson and Adamian. They present a numerical study of three Ritz–Galerkin approximation schemes (cubic B-splines, cubic Hermite splines, and natural modes) to compute functional feedback gains. These are obtained by combining optimal state estimator gains with state feedback control gains (computed using approximate solutions to the usual algebraic Riccati equation). No theory is given in the Gibson and Adamian paper since we had originally intended to include a separate contribution on the theory of the LQR problem. The appearance of the text [5] before this volume could be published made such a contribution less compelling.

Another topic in our original plans for this volume that was subsequently omitted included a treatment of theory and computation for identification and parameter estimation in distributed parameter systems. Again, appearance of a monograph [2] in the meantime greatly reduced the necessity for such an article.

The fourth paper in this volume, completed by Russell in 1987, is an overview discussion of damping models for flexible structures. As pointed out in the contributions of Delfour and Polis, Gibson and Adamian, and Curtain, the nature of damping in structures is of fundamental importance in the design of stabilizing controllers for flexible systems. Russell outlines several well-known beam damping models such as air or viscous damping, Kelvin–Voigt damping, and structural damping including the mathematically popular square root or $A^{1/2}$ damping operator. More importantly, Russell discusses at length what in subsequent years has come to be known as the spatial hysteresis damping model where the damping is due to differential rotation rates in adjacent segments of the beam. Such a model can be motivated by a fiber embedded composite material beam paradigm. It is of interest to note that in subsequent experiments and time domain identification procedures with composite material beams (see [1] for a summary of experimental and computational findings) using several of the damping models proposed here by Russell, it has been verified that spatial hysteresis is indeed a model of choice in describing quantitative and qualitative features of damping in certain composite material beams.

As we have noted above, our plans to include topics such as an LQR theory and parameter estimation for distributed parameter systems were not realized in this Frontiers volume. Fortunately, our attempts to include a contribution relating frequency domain techniques to state-space or time domain approaches for infinite-dimensional systems were more successful and resulted in a rather comprehensive and lucid presentation by Curtain. She focuses on these two approaches to control design in an integrated system theoretic framework. She discusses in some detail the so-called Callier–Desoer transfer function class and the Pritchard–Salamon class of state-space systems. As she amply demonstrates by examples, these classes include many infinite-dimensional systems of practical interest. In discussions of robustness, Curtain details results for robust (with respect to additive perturbations) stabilization control design for strictly proper systems in the Callier–Desoer class. In a companion presentation on exponentially stabilizable and detectable Pritchard–Salamon systems, the problem of robustness with respect to factor perturbation is discussed. Even though the topics of robust stabilization and H^∞ design on which Curtain focuses have experienced explosive growth and development since she completed her contribution in 1989, her article is highly recommended reading for researchers in both the frequency domain and time domain control communities. It epitomizes the type of contributions sought for this Frontiers series.

H. T. Banks
Raleigh, North Carolina
July 13, 1992

References

[1] H. T. Banks and D. J. Inman, *On damping mechanisms in beams*, ASME J. Appl. Mech., 58 (1991), pp. 716–723.

[2] H. T. Banks and K. Kunisch, *Estimation Techniques for Distributed Parameter Systems*, Birkhäuser–Boston, Boston, MA, 1989.

[3] W. Desch et al., Eds., *Estimation and Control of Distributed Parameter Systems*, Internat. Series Numer. Math., Vol. 100, Birkhäuser Verlag, Basel, Switzerland, 1991.

[4] F. Kappel et al., Eds., *Control and Estimation of Distributed Parameter Systems*, Internat. Series Numer. Math., Vol. 91, Birkhäuser Verlag, Basel, Switzerland, 1989.

[5] T. Lasiecka and R. Triggiani, *Differential and Algebraic Riccati Equations with Applications to Boundary/Point Control Problems: Continuous Theory and Approximation Theory*, Lecture Notes in Control and Info. Sci., Vol. 164, Springer-Verlag, Berlin, 1991.

Chapter 1

Pointwise Control for Distributed Systems

J. L. Lions

1.1. Introduction.

We consider in this chapter systems modeled by Partial Differential Equations of evolution, i.e., *evolution distributed systems.*

We suppose that we can act on the system only at some particular points, this is the so-called *pointwise control.*

This amounts, at least formally, to having a state equation given by

$$\frac{\partial y}{\partial t} + \mathbf{A}y = v(t)\delta(x-b) \tag{1.1}$$

where $\mathbf{A}$ is a linear partial differential operator, or a *system* of operators, and where $\delta(x-b)$ denotes the Dirac mass at a given point b—the point where we can act on the system.

In (1.1) the operator $\frac{\partial}{\partial t} + \mathbf{A}$ can be parabolic, hyperbolic or well set in Petrowsky's sense.

Of course, we add to (1.1) suitable initial and boundary conditions.

In (1.1) the control function v is generally subject to some constraints, which we express by

$$v \in \mathbf{U}_{\text{ad}}. \tag{1.2}$$

In some cases b can be on the boundary of the domain where we study (1.1). Then Dirac masses appear in the boundary conditions.

Of course, in applications, "points" are actually replaced by "small" regions around b—which amounts to replacing $\delta(x-b)$ in (1.1) by a function with "small" support around b. Then $\delta(x-b)$ can be thought of as a limiting case. But the pointwise model leads to questions to be addressed anyway, namely, what is the behaviour of the optimal control when the region around b gets smaller and smaller.

Assuming the state being given by the solution of (1.1) (subject to appropriate boundary and initial conditions), say $y = y(v)$, one defines next a *cost function*

(1.3) $$J(v) = \phi(y(v), v)$$

and one has to minimize $J(v)$, $v \in \mathbf{U}_{\mathrm{ad}}$.

One should be careful in defining $J(v)$. Indeed, we deal now with weak solutions cf. (1.1)—due to the presence of $\delta(x - b)$ in the right-hand side (or in the boundary conditions), and we need very precise information on the regularity of $y(v)$.

The "optimal" regularity results are generally obtained in the framework of Sobolev spaces. But here, if one uses the fact that

(1.4) $$\delta(x - b) \in H^{-\alpha}(\Omega) \quad \text{iff } \frac{1}{2} - \frac{\alpha}{n} < 0, \Omega \subset \mathbb{R}^n$$

one loses part of the information—in (1.4) $H^{-\alpha}(\Omega)$ denotes the dual space of $H_0^{\alpha}(\Omega)$; $H_0^{\alpha}(\Omega)$ is the closure in $H^{\alpha}(\Omega)$ of C^{∞} functions with compact support; $H^{\alpha}(\Omega)$ is the Sobolev space of functions that are in $L^2(\Omega)$ together with all their derivatives of "order α," these derivatives being defined, if α is not an integer, by Fourier transform or by interpolation; cf. Lions and Magenes [31] for precise definitions.

The general methodology that we propose is as follows:

(i) We define new function spaces such that

(1.5) $$J(v) < \infty.$$

For instance, let us assume that $\phi(y(v), v)$ is given by

(1.6) $$\phi(y(v), v) = \|Cy(v)\|_{\mathbf{H}} + \|v\|_{\mathbf{U}_1}$$

where

(1.7)
C maps the state $y(v)$ in a space $\tilde{\mathbf{H}}$
$\mathbf{H}$ and $\mathbf{U}_1$ are two Hilbert spaces, ($\| \ \|_X =$ norm in X),

and where we know that

(1.8) $$\mathbf{H} \subset \tilde{\mathbf{H}}.$$

Then we define a (possibly new) function space $\mathbf{U}$ by

(1.9) $$\mathbf{U} = \{v | v \in \mathbf{U}_1, Cy(v) \in \mathbf{H}\}.$$

We provide $\mathbf{U}$ with the norm of the graph

(1.10) $$\|v\|_{\mathbf{U}} = \left(\|v\|_{\mathbf{U}_1}^2 + \|Cy(v)\|_{\mathbf{H}}^2\right)^{1/2}$$

which makes $\mathbf{U}$ a Hilbert space.

(ii) Once the functional framework is defined as above, the next question is to characterize the space **U** in "classical" terms. In connection with this question, the following remarks are in order:

(a) A "classical" characterization may not exist, or may be just much more complicated than the direct definition (1.9).

Examples of such situations are given in §1.2, which is concerned with parabolic models.

(b) For hyperbolic systems, the above question leads to a large number of very interesting questions. Some of them are solved, others are still open. We refer to §1.3 and to the Bibliography.

(c) An optimality system can be derived, in a rather concrete way, in all situations, i.e., with **U** characterized or not in a "classical" fashion. General properties of **U** are sufficient, without an "explicit" definition[1].

These questions are studied in §1.2 for the diffusion equation and in §1.3 for the wave equation. Other models are considered in §1.4, where we address some other questions, always for pointwise controls, namely: (i) exact controllability and (ii) stabilizability.

All questions studied here arise in several contexts, in particular, for problems of §§1.3 and 1.4, in the control of flexible structures. Another interesting example has been mentioned to the author by H. T. Banks (cf. Banks, Propst, and Silcox [1]).

1.2. Diffusion Equation.

1.2.1. Setting of the Problem. Let Ω be a bounded open set of $\mathbb{R}^n$, with smooth boundary Γ.

The hypothesis that Ω is *bounded* can be relaxed in several situations, in the whole paper; but we shall assume Ω to be bounded to fix ideas and to simplify some questions. The same type of remark applies to the hypothesis that $\Gamma = \partial\Omega$ is *smooth*.

In the applications, $n = 1, 2, 3$, of course.

We consider an *evolution equation* in the cylinder

$$Q = \Omega \times]0, T[, \qquad T > 0, \text{ finite.} \tag{1.11}$$

We shall set

$$\Sigma = \Gamma \times]0, T[, \quad \text{the lateral boundary of } Q. \tag{1.12}$$

In this section, we consider a system with the state equation given by

[1] This is somewhat reminiscent of some questions arising in the theory of interpolation of function spaces.

(1.13) $$\frac{\partial y}{\partial t} - \Delta y = v(t)\delta(x-b) \quad \text{in } Q;$$

in (1.13):

$$\begin{aligned} \delta(x-b) &= \text{ Dirac mass at } b, \quad b \text{ given in } \Omega \\ v(t) &= \text{ control function (all functions are supposed real valued);} \end{aligned}$$

the initial condition and the boundary condition are given by

(1.14) $$y(x,0) = 0 \quad \text{in } \Omega$$

(1.15) $$y(x,t) = 0 \quad \text{on } \Sigma.$$

REMARK 1.2.1. *We could as well consider a nonzero initial condition in (1.14) and other boundary conditions on Σ; we shall return to this point.*

REMARK 1.2.2. *In a sense which we shall make precise below, the equations (1.13), (1.14), (1.15) (where it should be made precise in which space $v(t)$ lies) admit a unique solution, which is denoted by*

(1.16) $$y(x,t;v) \quad \textit{or} \quad y(v).$$

This is the state of the system.

In a formal manner (for the time being), we consider the *cost function* given by

(1.17) $$J(v) = \int_\Omega [y(x,T;v) - z_d]^2 \, dx + N \int_0^T v^2(t)dt$$

where

(1.18) $$\begin{aligned} z_d &\text{ is given in } L^2(\Omega) \\ N &\text{ is given } > 0. \end{aligned}$$

In order for (1.7) to make sense, a *necessary* condition is that

(1.19) $$v \in L^2(0,T).$$

The (formal) problem we are interested in consists in finding

(1.20) $$\inf J(v)$$

when v spans a given subset of $L^2(0,T)$.

1.2.2. The Weak Solution of the State Equation. We define, as it is by now standard (cf. Lions and Magenes [31]), the weak solution of (1.3), (1.4), (1.5) by transposition.

Let ψ be given in Q, smooth (this being made precise later on) and let ϕ be the solution of

$$
\begin{aligned}
-\frac{\partial\phi}{\partial t} - \Delta\phi = \psi \quad & \text{in } Q \\
\phi(x,T) = 0 \quad & \text{in } \Omega, \\
\phi = 0 \quad & \text{on } \Sigma.
\end{aligned} \tag{1.21}
$$

We shall write $\phi', y', \cdots$ for $\partial\phi/\partial t, \partial y/\partial t, \cdots$.

If we multiply (1.13) by ϕ, integrate on Q, and integrate by parts—in a formal manner!—we obtain

$$
\int_0^T \phi(b,t)v(t)dt = \int_Q y\psi \, dx \, dt. \tag{1.22}
$$

We now take (1.22) as the definition of $y = y(v)$:

$$
\int_Q y(v)\psi \, dx \, dt = \int_0^T v(t)\phi(b,t) \, dt, \tag{1.23}
$$

where ϕ is the solution of (1.11).

If we assume that

$$
\psi \in L^2(0,T; H_0^\alpha(\Omega)) \tag{1.24}
$$

where

$$
H^\alpha(\Omega) = \text{ Sobolev space}^2 \text{ of order } \alpha \text{ on } \Omega \tag{1.25}
$$

and

$$
\begin{aligned}
H_0^\alpha(\Omega) = \; & \text{closure in } H^\alpha(\Omega) \text{ of the space } D(\Omega) \text{ of} \\
& C^\infty \text{ functions with compact support,}
\end{aligned} \tag{1.26}
$$

then, for α large enough (depending on the dimension n), the mapping

$$
\psi \to \phi(b,t)
$$

is linear and continuous from $L^2(0,T; H_0^\alpha(\Omega)) \to L^2(0,T)$; therefore, given $v \in L^2(0,T)$, the mapping

$$
\psi \to \int_0^T v(t)\phi(b,t)dt
$$

[2] This is the classical space if α = integer. For α not an integer it can be defined either by interpolation theory or by Fourier transform.

is continuous from $L^2(0,T;H_0^\alpha(\Omega)) \to \mathbb{R}$, so that (1.23) uniquely defines $y(v) \in L^2(0,T;H^{-\alpha}(\Omega))[H^{-\alpha}(\Omega) = \text{dual of } H_0^\alpha(\Omega)]$.

If $n = 2$ or 3, we can take $\alpha = 0$, hence

$$y(v) \in L^2(Q). \tag{1.27}$$

The state of the system is now defined by (1.23).

REMARK 1.2.3. *Let us assume—to simplify—that the spectrum is simple: if we introduce*

$$-\Delta w_j = \lambda_j w_j, \quad w_j = 0 \quad \text{on } \partial\Omega, \tag{1.28}$$

we assume that all eigenvalues are simple

$$0 < \lambda_1 < \lambda_2 < \cdots.$$

Then, $y(v)$ as defined by (1.23), *is given by*

$$y(v) = \sum_j \left(\int_0^t e^{-\lambda_j(t-\sigma)} v(\sigma) d\sigma \right) w_j(b) w_j \tag{1.29}$$

where convergence takes place in $L^2(Q)$ (if $n \leq 3$).

REMARK 1.2.4. *Except in the case where $n = 1$, there are functions v in $L^2(0,T)$ such that $y(.,t;v)$ is* not *continuous from $[0,T] \to L^2(\Omega)$; in other words, there are elements $v \in L^2(0,T)$ such that*

$$J(v) = +\infty.$$

(For technical details, cf., for instance, Lions, [20], [21] *and cf. also* §1.2.3 *below.)*

At this point, it is natural to introduce the space of elements $v \in L^2(0,T)$ such that $J(v) < \infty$. This is the goal of §1.2.3.

1.2.3. The Space U. We define the space $\mathbf{U}$ as follows:

$$\mathbf{U} = \{v | v \in L^2(0,T), y(T;v) \in L^2(\Omega)\}. \tag{1.30}$$

REMARK 1.2.5. *It can easily be verified that, although $t \to y(t;v)$ is* not *continuous from $[0,T] \to L^2(\Omega)$, it is continuous from $[0,T] \to H^{-\beta}(\Omega)$, β large enough. Therefore* (1.30) *makes sense.*

We provide $\mathbf{U}$ with the norm

$$\|v\|_{\mathbf{U}} = \left[\int_0^T v^2 dt + \int_\Omega y(x,T;v)^2 dx \right]^{1/2}; \tag{1.31}$$

with this norm, $\mathbf{U}$ becomes a Hilbert space (on $\mathbb{R}$).

The following properties are quite easy to verify (cf. Lions [21, p. 241ff], for technical details):

(1.32) **U** does not depend on $b \in \Omega$ (up to an equivalence of norm);

(1.33) **U** does not depend on Ω, again up to an equivalence of norm,

and does not depend on the boundary conditions.

Therefore **U** coincides with the space of v's $\in L^2(0,T)$ such that

$$\phi(T;v) \in L^2(\mathbb{R}^n) \tag{1.34}$$

where ϕ is the solution of the Cauchy problem:

$$\begin{aligned} &\phi' - \Delta\phi = v(t)\ \delta(x) \quad \text{in } \mathbb{R}^n x]0,T[, \\ &\phi(x,0) = 0. \end{aligned} \tag{1.35}$$

Of course, one has an explicit formula for the solution ϕ of (1.34). Expressing that (1.34) holds true leads—in a *formal* way—to the condition:

$$\int_0^T \int_0^T (2T - (t+s))^{-n/2} v(t)v(s)dt\,ds < \infty. \tag{1.36}$$

REMARK 1.2.6. *Condition* (1.36) *is essentially correct, and it is correct if* $v > 0$; *one has to take care of the commutation in the integrations by parts leading to* (1.36); *cf. Simon* [42] *for precise results on this question.*

We can now state, in a precise form, the optimal control problem we want to solve: we are given

$$\mathbf{U}_{\text{ad}} = \text{ closed convex subset of } \mathbf{U}. \tag{1.37}$$

Find

$$\inf J(v) \quad \forall v \in \mathbf{U}_{\text{ad}}. \tag{1.38}$$

It is obvious that problem (1.38) admits a unique solution, since

$$J(v) \geq c_0 \|v\|^2_{\mathbf{U}} \quad \text{as } \|v\| \to +\infty, \tag{1.39}$$

and since $v \to J(v)$ is strictly convex.

Let $u \in \mathbf{U}_{\text{ad}}$ be this unique solution:

$$J(u) = \min J(v) \quad \forall v \in \mathbf{U}_{\text{ad}}. \tag{1.40}$$

We want now to characterize u through the solution of the optimality system (O.S.).

1.2.4. Optimality System. It can be verified that

$$\mathbf{D}(]0,T[) \text{ is dense in } \mathbf{U}. \tag{1.41}$$

We can then introduce $\mathbf{U}'$ = dual space of $\mathbf{U}$, as a space of distributions on $]0,T[$; we have

$$\mathbf{U} \subset L^2(0,T) \subset \mathbf{U}'. \tag{1.42}$$

Then the optimal control u, solution of (1.40), is characterized by the unique solution $\{u, y, p\}$ of the O.S.

$$y' - \Delta y = u\sigma(x-b), \tag{1.43a}$$

$$-p' - \Delta p = 0, \tag{1.43b}$$

$$y(x,0) = 0, \qquad p(x,T) = y(x,T) - z_d(x), \tag{1.43c}$$

$$y = p = 0 \quad \text{on } \Sigma, \tag{1.43d}$$

$$\int_0^T (p(b,t) + Nu(t))(v(t) - u(t))dt \geq 0 \quad \forall v \in \mathbf{U}_{\text{ad}}. \tag{1.44}$$

REMARK 1.2.7. *One can verify that if p is defined as the solution of* (1.43b), *with $p(x,T)$ given in $L^2(\Omega)$, and with the boundary condition $p = 0$ on Σ (the same property holds true for other coundary conditions), then*

$$p(b,t) \in \mathbf{U}'. \tag{1.45}$$

Therefore, in (1.44), *the integral $\int_0^T p(b,t)(v(t) - u(t))dt$ expresses the duality between $\mathbf{U}'$ and $\mathbf{U}$.*

1.2.5. Examples.

Example 1.2.1. $\mathbf{U}_{\text{ad}} = \mathbf{U}$. Then (1.44) reduces to

$$p(b,t) + Nu(t) = 0. \tag{1.46}$$

Using (1.46) in (1.43a) leads to a system in $\{y, p\}$, which can be uncoupled leading to an infinite-dimensional Riccati partial differential equation; cf. Lions [21].

Example 1.2.2. $\mathbf{U}_{\text{ad}} = \{v | v \in \mathbf{U}, v \geq 0 \text{ a.e. in }]0,T[\}$.

Then

$$u(t) = \frac{1}{N} p(b,t)^- \tag{1.47}$$

(where $\lambda^- = \sup(-\lambda, 0)$).

Example 1.2.3. $\mathbf{U}_{\text{ad}} = \{v | v \in \mathbf{U},\ 0 \leq v \leq M \text{ a.e. in }]0,T[\}$. Let us show a bang-bang property.

We shall say that b is a strategic point[3] if all eigenvalues λ_j in (1.28) are simple and if

$$w_j(b) \neq 0 \quad \forall j. \tag{1.48}$$

We remark that strategic points are *dense* in Ω. Let us check the following:

$$\left| \begin{array}{l} \text{if } b \text{ is strategic, if } z_d \text{ is not } C^\infty \text{ outside } b, \text{ and} \\ \text{if } \mathbf{U}_{\text{ad}} \text{ is given as above,} \\ \text{then } u(t) = 0 \text{ or } = M \text{ a.e. in }]0,T[. \end{array} \right. \tag{1.49}$$

Indeed, $p(b,t)$ is given by

$$p(b,t) = \sum_j e^{-j(T-t)}(y(T) - z_d, w_j)w_j(b). \tag{1.50}$$

If $0 < u(t) < M$ on a set E of positive measure on $]0,T[$, then $p(b,t) = 0$ on E, and since $p(b,t)$ can be analytically extended in $t, p(b,t) = 0$ in $]0,T[$, and then

$$(y(T) - z_d, w_j)w_j(b) = 0 \quad \forall j.$$

Since $w_j(b) \neq 0\,\forall j$, it follows that $(y(T) - z_d, w_j) = 0\,\forall j$, hence $y(T) - z_d = 0$. But this is impossible if z_d is not C^∞ outside b, hence (1.49) follows.

REMARK 1.2.8. *If $\Omega =]0,2\pi[$, b "strategic" corresponds to b irrational—certainly a strange condition! But one has to remember that, in the applications and in the numerical computations, one deals with a finite number of eigenvalues and eigenfunctions. Then condition* (1.48) *becomes reasonable!*

1.2.6. Approximation of the Dirac Mass. Let us introduce

$$\begin{array}{l} \chi_\varepsilon = \text{function} \in L^2(\Omega), \quad \chi_\varepsilon \geq 0, \quad \displaystyle\int_\Omega \chi_\varepsilon \, dx = 1, \\ \chi_\varepsilon \text{ with support in the ball of center } b \\ \text{and radius } \varepsilon \ (\varepsilon \text{ small enough}). \end{array} \tag{1.51}$$

If we consider the state equation given by

$$\begin{array}{l} y' - \Delta y = v(t)\chi_\varepsilon \\ y(x,0) = 0, \quad y = 0 \quad \text{on } \Sigma, \end{array} \tag{1.52}$$

it admits a unique solution, which is continuous from $[0,T] \rightarrow L^2(\Omega)$. Let us denote by $y_\varepsilon(v)$ this solution. If we define

[3]This notation has been used by El Jai [5] in another setting; cf. also the book by El Jai and Pritchard [6].

(1.53) $$J_\varepsilon(v) = \int_\Omega [y_\varepsilon(x,T;v) - z_d(x)]^2\,dx + N\int_0^T v^2 dt,$$

then $v \to J_\varepsilon(v)$ is continuous on $L^2(0,T)$ (there is no need to introduce $\mathbf{U}$ this time!). Let us consider the problem

(1.54) $$\inf J_\varepsilon(v), \quad v \in \mathbf{U}^1_{\mathrm{ad}}$$

(1.55) $$\mathbf{U}^1_{\mathrm{ad}} = \text{ closed convex subset of } L^2(0,T).$$

This problem admits a unique solution u_ε, which is characterized by the O.S.

(1.56) $$\begin{aligned} y'_\varepsilon - \Delta y_\varepsilon &= u_\varepsilon \chi_\varepsilon, \\ -p'_\varepsilon - \Delta p_\varepsilon &= 0, \\ y_\varepsilon(0) &= 0, \qquad p_\varepsilon(T) = y_\varepsilon(T) - z_d, \\ y_\varepsilon = p_\varepsilon &= 0 \quad \text{on } \Sigma \end{aligned}$$

and

(1.57) $$\int_0^T \left[\left(\int_\Omega p_\varepsilon \chi_\varepsilon dx\right) + N u_\varepsilon\right](v(t) - u_\varepsilon(t))dt \geq 0 \quad \forall v \in \mathbf{U}^1_{\mathrm{ad}}.$$

Let us see what happens as $\varepsilon \to 0$. We introduce

(1.58) $$\mathbf{U}_{\mathrm{ad}} = \text{ closure in } \mathbf{U} \text{ of } \mathbf{U}^1_{\mathrm{ad}} \cap \mathbf{U},$$

assuming that $\mathbf{U}^1_{\mathrm{ad}} \cap \mathbf{U} \neq \phi$.

Let u be the solution of (1.40). Then, as $\varepsilon \to 0$,

(1.59) $$u_\varepsilon \to u \quad \text{in } L^2(0,T).$$

1.2.7. Dual Problem. We can apply duality theory, in the framework of Rockafellar [38] and Ekeland and Temam [7], to problem (1.40).

We define

(1.60) $$F(v) = N\int_0^T v^2 dt, \qquad v \in L^2(0,T),$$

(1.61) $$G(q) = \int_\Omega (q - z_d)^2 dx, \qquad q \in L^2(\Omega),$$

(1.62) $$Lv = y(T;v), \qquad D(L)(\text{domain of } L) = \mathbf{U}.$$

If $\mathbf{U}_{\mathrm{ad}} = \mathbf{U}$, problem (1.40) becomes

(1.63) $$\inf[F(v) + G(Lv)].$$

If F^* (respectively, G^*) denotes the conjugate functional of F (respectively, G) (for instance, $F^*(v) = \sup_w[(v,w) - F(w)]$), then

$$\inf_{v\in D(L)} [F(v) + G(Lv)] = - \inf_{q\in D(L^*)} [F^*(L^*q)) + G^*(-q)] \tag{1.64}$$

where L^* is the adjoint of L, with domain L^*.

We refer to Lions [21, Chap. 4, §3], for the (easy) details of computations. It leads to the following class of problems.

The state equation is now given by

$$\begin{aligned} &y' - \Delta y = 0 \quad \text{in } Q, \\ &y(x,0) = v(x) \quad \text{in } \Omega, \\ &y = 0 \quad \text{on } \Sigma. \end{aligned} \tag{1.65}$$

We take $v \in L^2(\Omega)$ in (1.65). The solution is denoted by $y(x,t;v) = y(v)$. The cost function is given now by

$$J(v) = \int_0^T [y(b,t;v) - z_d(t)]^2\, dt + N \int_\Omega v^2 dx, \tag{1.66}$$

where z_d is given in $L^2(0,T)$ and where N is given > 0.

Since $y(v)$ is C^∞ in x, t for $t > 0$, $y(b,t;v)$ makes sense—but for $v \in L^2(\Omega), y(b,t;v)$ does not necessarily belong to $L^2(0,T)$.

This leads to the following: we introduce a new space $\mathbf{H}$ by

$$\mathbf{H} = \{v | v \in L^2(\Omega), y(b,t;v) \in L^2(0,T)\}. \tag{1.67}$$

Provided with the norm

$$\|v\|_{\mathbf{H}} = \left(\int_\Omega v^2 dx + \int_0^T y(b,t;v)^2 dt \right)^{1/2}, \tag{1.68}$$

$\mathbf{H}$ becomes a Hilbert space.

If we assume that

$$\mathbf{U}_{\text{ad}} = \text{ closed convex subset of } \mathbf{H}, \tag{1.69}$$

we are interested in finding

$$\inf J(v), \quad v \in \mathbf{U}_{\text{ad}}. \tag{1.70}$$

This problem admits a unique solution u.

The optimal control u is characterized by the O.S.

$$
\begin{aligned}
&y' - \Delta y = 0,\\
&-p' - \Delta p = (y(b,t) - z_d(t))\delta(x - b),\\
&y = p = 0 \quad \text{on } \Sigma,\\
&y(0) = 0, \qquad p(T) = 0,
\end{aligned}
\tag{1.71}
$$

and

$$
\int_\Omega (p(x,0) + Nu(x))(v(x) - u(x))dx \geq 0 \quad \forall v \in \mathbf{U}_{\mathrm{ad}}. \tag{1.72}
$$

REMARK 1.2.9. *We have*

$$
p(x,0) \in \mathbf{H}', \qquad \mathbf{H} \subset L^2(\Omega) \subset \mathbf{H}'; \tag{1.73}
$$

in (1.72), $\int_\Omega p(x,0)(v(x) - u(x))dx$ *expresses the duality between* $\mathbf{H}'$ *and* $\mathbf{H}$.

REMARK 1.2.10. *Let us assume that*

$$
\mathbf{U}_{\mathrm{ad}} = \{v | v \in \mathbf{U}, \quad 0 \leq v \leq M\}. \tag{1.74}
$$

Then one has the bang-bang property:

$$
u = 0 \quad \text{or} \quad u = M \quad \text{a.e. in } \Omega. \tag{1.75}
$$

This follows from (1.72) *and from the fact that* $p(x,0)$ *cannot vanish on a set of* > 0 *measure, except if* $y(b,t) = z_d(t)$*—a situation which is excluded if we assume that* z_d *is* not C^∞ *for* $t > 0$.

1.2.8. Various Remarks. In what has been said, $-\Delta$ can be replaced by A, given by

$$
A = -\frac{\partial}{\partial x_i}\left(a_{ij}(x,t)\frac{\partial}{\partial x_j}\right) \tag{1.76}
$$

(we use the summation convention of repeated indices), where

$$
a_{ij}(x,t)\xi_i\xi_j \geq \alpha\xi_i\xi_j, \qquad \alpha > 0. \tag{1.77}
$$

If we define $y(v)$ as the solution of

$$
\begin{aligned}
&y'(v) + Ay(v) = v(t)\delta(x - b),\\
&y(x,0;v) = 0,\\
&y(x,t;v) = 0 \quad \text{on } \Sigma,
\end{aligned}
\tag{1.78}
$$

then we can define

$$
\mathbf{U}_A = \{v | v \in L^2(0,T), y(T;v) \in L^2(\Omega)\}, \tag{1.79}
$$

provided with the graph norm as before.

If the a_{ij}'s are smooth enough[4], then

$$\mathbf{U}_A = \mathbf{U} \tag{1.80}$$

up to an equivalence of norm.

REMARK 1.2.11. *We refer to Shih* [41] *and Shih and Simon* [42] *for similar problems in situations where b actually depends on $t : b = b(t)$.*

REMARK 1.2.12. *It is a simple matter to give nonlinear extensions of the situation of* 1.2.6. *As an example, let us consider the state equation*

$$\begin{aligned} &y' - \Delta y + y^3 = v(t)\chi_\varepsilon \quad in\, Q, \\ &y(x,0) = 0, \qquad y = 0 \quad on\, \Sigma. \end{aligned} \tag{1.81}$$

Problem (1.81) admits a unique solution, denoted by $y(v)$. We can consider the cost function (1.53), where $y(v)$ is now given by the solution of (1.81).

The problem inf $J_\varepsilon(v)$ admits a solution, which is not necessarily unique (we lose the convexity). For every optimal control u_ε there is an O.S.:

$$\begin{aligned} &y_\varepsilon' - \Delta y_\varepsilon + y_\varepsilon^3 = u_\varepsilon \chi_\varepsilon, \\ &-p_\varepsilon' - \Delta p_\varepsilon + 3y_\varepsilon^2 p_\varepsilon = 0, \\ &y_\varepsilon(0) = 0, \qquad p_\varepsilon(T) = y_\varepsilon(T) - z_d, \\ &y_\varepsilon = p_\varepsilon = 0 \quad \text{on } \Sigma. \end{aligned} \tag{1.82}$$

But it does not seem possible to let $\varepsilon \to 0$ in the above problem.

REMARK 1.2.13. *All that has been said extends to systems of parabolic equations (of course, with new spaces, constructed in the same way as* **U** *has been constructed above).*

1.3. Wave Equation.

1.3.1. Setting of the Problem. We consider now the *pointwise control* of the wave equation. The state equation is given by

$$y'' - \Delta y = v(t)\delta(x-b), \tag{1.83a}$$

$$y(x,0) = 0, \tag{1.83b}$$

$$y'(x,0) = 0 \quad \text{in } \Omega, \tag{1.83c}$$

$$y = 0 \quad \text{on } \Sigma, \tag{1.83d}$$

where b is given in Ω.

[4] $a_{ij}, \partial a_{ij}/\partial t, \partial^2 a_{ij}/(\partial x_k \partial x_l) \in L^\infty(Q)\ \forall k,l$; cf. Li Ta Tsien [32], [33] for technical details. Cf. also, in J. L. Lions [21, Chap. 4, Appendices 1 and 2], written by Li Ta Tsien.

This problem admits a unique weak solution, which is denoted by $y(v)$. This will be made precise in §1.3.2 below.

Let us admit for the time being that we can define

$$y(.,T;v).$$

We will show later on that if $n \leq 3, y(.,T;v) \in L^2(\Omega)$. We then consider the cost function

$$J(v) = \int_\Omega [y(x,T;v) - z_d(x)]^2 \, dx + N \int_0^T v^2 dt, \tag{1.84}$$

where

$$z_d \text{ is given in } L^2(\Omega),$$

$$N \text{ is given } > 0.$$

Let us introduce

$$\mathbf{U}_{\text{ad}} = \text{ closed convex subset of } L^2(0,T). \tag{1.85}$$

We are looking for

$$\inf J(v), \quad v \in \mathbf{U}_{\text{ad}}. \tag{1.86}$$

We will show that this problem admits a unique solution u, and we will give the O.S. that characterizes u.

1.3.2. The Weak Solution of the State Equation. We proceed as in §1.2.2 by duality. Let ψ be given, smooth enough for the time being. Let ψ be the solution of

$$\begin{aligned} &\phi'' - \Delta\phi = \psi \quad \text{in } Q,\\ &\phi(T) = \phi'(T) = 0 \quad \text{in } \Omega,\\ &\phi = 0 \quad \text{on } \Sigma. \end{aligned} \tag{1.87}$$

Then, if we multiply (1.83a) by ϕ and if we integrate formally by parts, we obtain

$$\int_Q y(v)\psi \, dx \, dt = \int_0^T \phi(b,t)v(t)dt. \tag{1.88}$$

We then use (1.88) for defining $y(v)$.

If we take

$$\psi \in L^1(0,T;H_0^1(\Omega)) \tag{1.89}$$

then, as it is classical,

$$\phi \text{ is continuous from } [0,T] \to H^2(\Omega). \tag{1.90}$$

If we assume that

$$n \leq 3 \tag{1.91}$$

it follows from (1.90) and from the fact that

$$H^2(\Omega) \subset C^\circ(\Omega) = \text{ continuous functions in } \Omega \tag{1.92}$$

that

$$\phi(b,t) \text{ is continuous in } [0,T]. \tag{1.93}$$

Therefore, we have in particular that

$$\psi \to \phi(b,t)$$

is a continuous linear mapping from $L^1(0,T;H_0^1(\Omega)) \to L^2(0,T)$. Therefore (1.88) uniquely defines

$$y(v) \in L^\infty(0,T;H^{-1}(\Omega)) \tag{1.94}$$

as the solution (in a weak sense) of (1.83).

But the above are very crude estimates. We are now going to show (by several methods) that

$$\begin{aligned} &\text{if } v \in L^2(0,T) \text{ and if } n \leq 3, \\ &t \to y(t;v) \text{ is continuous from } [0,T] \to L^2(\Omega). \end{aligned} \tag{1.95}$$

1.3.3. Continuity in $L^2(\Omega)$. (Part I.) Let us assume that $n = 3$.

It is a very natural thing to compare the solution of (1.83) with the solution of the Cauchy problem

$$\begin{aligned} &\phi'' - \Delta\phi = v(t)\delta(x-b), \\ &\phi(0) = \phi'(0) = 0 \quad \text{in } \mathbb{R}^n. \end{aligned} \tag{1.96}$$

There is no restriction in assuming that $b = 0 \in \Omega$. Then ϕ is explicitly given by

$$\phi = \frac{1}{4\pi|x|}v(t-|x|), \qquad |x| \leq t. \tag{1.97}$$

If we set

$$z = y - \phi, \tag{1.98}$$

$$\begin{aligned} &z'' - \Delta z = 0 \quad \text{in } Q, \\ &z(0) = z'(0) = 0, \\ &z = -\phi \quad \text{on } \Sigma. \end{aligned} \tag{1.99}$$

But if $v \in L^2(0,T)$, then the trace $\phi|_\Sigma$ of ϕ satisfies

$$\phi|_\Sigma \in L^2(\Sigma) \tag{1.100}$$

and there is a constant c such that

$$\|\phi|_\Sigma\|_{L^2(\Sigma)} \leq C \left(\frac{T}{d(b,\Gamma)}\right)^{1/2} \|v\|_{L^2(0,T)} \tag{1.101}$$

where $d(b,\Gamma)$ equals the distance from b to Γ.

It remains to solve (1.99) with ϕ satisfying (1.100).

This last question is of independent interest: changing notation, we want to find a function w solution of

$$\begin{aligned} &w'' - \Delta w = 0, \\ &w(0) = w'(0) = 0, \\ &w = g \quad \text{on } \Sigma, \end{aligned} \tag{1.102}$$

where

$$g \in L^2(\Sigma). \tag{1.103}$$

Let us admit for a moment that

(1.104) problem (1.102) admits a unique solution, which is continuous from $[0,T] \to L^2(\Omega)$.

Then (1.95) follows from (1.98), (1.100), and (1.104).

Proof of (1.104). It is again a proof by duality. Let ϕ be the solution of ψ, where ψ is given in $L^1(0,T;L^2(\Omega))$. Then, formally, if w is solution of (1.102), one has

$$\int_Q w\psi \, dx \, dt = -\int_\Sigma g \frac{\partial \phi}{\partial \nu} d\Sigma. \tag{1.105}$$

But (cf. Lions [23, Chap. 2, §4]), there is a constant c such that

$$\left\| \frac{\partial \phi}{\partial \nu} \right\|_{L^2(\Sigma)} \leq c\|\psi\|_{L^1(0,T;L^2(\Omega))} \tag{1.106}$$

and therefore $w \in L^\infty(0,T;L^2(\Omega))$, with

$$\|w\|_{L^\infty(0,T;L^2(\Omega))} \leq c\|g\|_{L^2(\Sigma)}. \tag{1.107}$$

But if g is smooth, w is smooth and in particular

$$w \in C^0([0,T]; L^2(\Omega)) = \text{ space of continuous function from } [0,T] \to L^2(\Omega). \tag{1.108}$$

Then (1.104) follows (1.107) and (1.108).

REMARK 1.3.1. *The above proof of* (1.95) *has been given in Lions* [24]. *It extends to the case when* $-\Delta$ *is replaced by* A *given by*

$$A = -\frac{\partial}{\partial x_i}\left(a_{ij}(x,t)\frac{\partial}{\partial x_j}\right), \tag{1.109}$$

$$\begin{aligned} &a_{ij} = a_{ji} \in L^\infty(Q), && \frac{\partial a_{ij}}{\partial t} \in L^\infty(Q), \\ &a_{ij}(x,t)\xi_i\xi_j \geq \alpha\xi_i\xi_i, && \alpha > 0. \end{aligned} \tag{1.110}$$

But it does not seem to extend (at least in a simple way!) to the case when the boundary condition (of the Dirichlet type) $y = 0$ on Σ is replaced by a boundary condition of the Neumann type $\partial y/\partial\nu = 0$ on Σ. This is related to the "optimum" regularity results for nonhomogeneous boundary conditions of the Neumann type for the wave equation; cf. Lasiecka and Triggiani [17].

REMARK 1.3.2. *For a systematic study of nonhomogeneous boundary conditions of the Dirichlet type—and the "optimum" regularity results—we refer to Lasiecka, Lions, and Triggiani* [19].

1.3.4. Continuity in $L^2(\Omega)$. (Part II.) Another proof of (1.95) has been given in Nirenberg [37]. If we consider the solution ϕ of

$$\left|\begin{aligned} &\phi'' - \Delta\phi = 0, \\ &\phi(T) = 0, \qquad \phi'(T) = h \\ &\phi = 0 \quad \text{on } \Sigma \end{aligned}\right. \tag{1.111}$$

then (assuming h to be smooth enough), we have

$$\int_\Omega y(x,T;v)h(x)dx = -\int_0^T \phi(b,t)v(t)dt \tag{1.112}$$

and we shall have the desired property (1.95) if we show that

$$h \to \phi(b,t) \text{ is continuous from } L^2(\Omega) \to L^2(0,T). \tag{1.113}$$

As a matter of convenience, we reverse time. Therefore we have to consider the solution ϕ of the equation

$$\begin{aligned} &\phi'' - \Delta\phi = 0, \\ &\phi(0) = 0, \qquad \phi'(0) = h, \\ &\phi = 0 \quad \text{in } \Sigma. \end{aligned} \tag{1.114}$$

Following Nirenberg [37], we verify that

$$\int_0^T \phi(b,t)^2 dt \le \frac{T}{2\pi d(b,\Gamma)} \int_\Omega h^2(x)dx, \qquad \Omega \subset \mathbb{R}^3. \tag{1.115}$$

The proof is based on the *finite speed of propagation.*

If we consider the Cauchy problem

$$\begin{gathered} \phi'' - \Delta\phi = 0 \quad \text{in } \mathbb{R}^3 \times \{t > 0\} \\ \phi(x,0) = \phi^0(x), \qquad \phi'(x,0) = \phi'(x) \end{gathered} \tag{1.116}$$

then according to a classical formula

$$\phi(0,t) = \frac{t}{4\pi} \int_{|w|=1} \left[\phi^1(tw) + \sum \frac{\partial \phi^0}{\partial x_i}(tw) w_i \right] dw \tag{1.117}$$

(we take $b = 0$). Therefore

$$\phi(0,t)^2 \le \frac{t^2}{2\pi} \int_{|w|=1} \left[\phi^1(tw)^2 + \sum \left(\frac{\partial \phi^0}{\partial x_i} \right)^2 (tw) \right];$$

hence

$$\int_0^{t_0} \phi(0,t)^2 dt \le \frac{1}{2\pi} \int_{|x| \le t_0} \left[\phi^1(x)^2 + \sum \left(\frac{\partial \phi^0}{\partial x_i} \right)^2 (x) \right] dx \tag{1.118}$$

or

$$\int_0^{t_0} \phi(0,t)^2 dt \le \frac{1}{2\pi} \int_{|x| \le t_0} \left[\left| \frac{\partial \phi}{\partial t}(x,0) \right|^2 + \sum \left| \frac{\partial \phi}{\partial x_i}(x,0) \right|^2 \right] dx. \tag{1.119}$$

Let us now divide $(0,T)$ in intervals $(0,d)$, $(d,2d)$, $\cdots, d = d(b,\Gamma), b = 0$ until $((N-1)d, Nd), Nd \ge T$.

We can apply (1.119) to ϕ by virtue of the finite speed of propagation:

$$\begin{aligned} \int_0^d \phi^2(0,t)dt &\le \frac{1}{2\pi} \int_{|x|<d} \left[\left| \left(\frac{\partial \phi}{\partial t}(x,0) \right) \right|^2 + \sum \left(\frac{\partial \phi}{\partial x_i}(x,0) \right)^2 \right] dx \\ &\le \frac{1}{2\pi} \int_\Omega \left[\phi'(0)^2 + \sum \left(\frac{\partial \phi}{\partial x_i}(0) \right)^2 \right] dx; \end{aligned} \tag{1.120}$$

$$\begin{aligned} \int_d^{2d} \phi^2(0,t)dt &\le \frac{1}{2\pi} \int_{|x| \le d} \left[\frac{\partial \phi}{\partial t}(x,d)^2 + \sum \frac{\partial \phi}{\partial x_i}(x,d)^2 \right] dx \\ &\le \frac{1}{2\pi} \int_\Omega \left[\frac{\partial \phi}{\partial t}(x,d)^2 + \sum \frac{\partial \phi}{\partial x_i}(x,d)^2 \right] dx, \cdots. \end{aligned} \tag{1.121}$$

But

$$\int_\Omega \left[\frac{\partial\phi}{\partial t}(x,t)^2 + \sum \frac{\partial\phi}{\partial x_i}(x,t)^2\right] dx = \int_\Omega h^2(x)dx.$$

Adding up the inequalities (1.120), (1.121), $\cdots$, we obtain

$$\int_0^{Nd} \phi^2(0,t)dt \leq \frac{N}{2\pi}\int h^2(x)dx,$$

hence (1.115) follows.

REMARK 1.3.3. *The above proof is independent of the type of boundary conditions (provided they are of local type, with a finite speed of propagation).*

1.3.5. Continuity in $L^2(\Omega)$. (Part III.) Another proof of (1.95) uses harmonic analysis. The proof is due to Meyer [34], [35] and it is actually the first proof which has been given of (1.95)!

One proves (1.115) (actually with a slightly less precise constant than $1/2\pi$ in (1.115)), using the explicit solution

$$\phi = \Sigma_j (h, w_j) w_j \frac{\sin t\sqrt{\lambda j}}{\sqrt{\lambda j}} \tag{1.122}$$

(with the notation of (1.28)).

One uses the next results on nonperiodic trigonometric sums (cf. Meyer [34]–[36]).

REMARK 1.3.4. *This proof seems to extend to equations of the following type:*

$$\begin{aligned} &\frac{\partial^2 y}{\partial t^2} + \Delta^2 y = v(t)\delta(x-b) \quad \text{in } Q,\\ &y(0) = y'(0) = 0,\\ &y = \frac{\partial y}{\partial v} = 0 \quad \text{on } \Sigma. \end{aligned} \tag{1.123}$$

It seems likely that, if $\Omega \subset \mathbb{R}^3$ and if $v \in L^2(0,T)$, then the solution y of (1.123) is continuous from $[0,T] \to H^{3/2}(\Omega)$.

The method of §1.1.4 *does not apply—since we have an infinite speed of propagation in the present equation—the method of* §1.1.3 *could apply, but a complete proof is not known to the author at the time of writing this paper.*

Note Added in Proof. (1) Cf. S. Jaffard, dissertation, Paris IX, April 1992. (2) For still another proof, cf. recent work of W. G. Pritchard.

1.3.6. Optimality System. Let us return now to problem (1.86). Since

$$v \to y(.,T;v)$$

is continuous from $L^2(0,T) \to L^2(\Omega)$, problem (1.86) admits a unique solution u.

The optimal control u is characterized by the solution $\{u, y, p\}$ of the optimality system (O.S.) given by

$$(1.124)\qquad \begin{aligned} &y'' - \Delta y = u\delta(x-b),\\ &p'' - \Delta p = 0;\\ &y(0) = y'(0) = 0, \quad p(T) = 0, \quad p'(T) = -(y(T) - z_d),\\ &y = p = 0 \quad \text{on } \Sigma, \end{aligned}$$

and

$$(1.125)\qquad \int_0^T (p(b,t) + Nu(t))(v(t) - u(t)) \geq 0 \quad \forall v \in \mathbf{U}_{\mathrm{ad}}, \quad u \in \mathbf{U}_{\mathrm{ad}}.$$

REMARK 1.3.5. *In* (1.125) *we use the fact that* $p(b,t) \in L^2(0,T)$.

1.3.7. Examples.

Example 1.3.1. If $\mathbf{U}_{\mathrm{ad}} = L^2(0,T)$, then (1.125) reduces to

$$(1.126)\qquad p(b,t) + Nu(t) = 0 \quad \text{in }]0,T[.$$

Example 1.3.2 We assume now that

$$(1.127)\qquad \mathbf{U}_{\mathrm{ad}} = \{v | \quad 0 \leq v \leq M\}.$$

We observe that

$$(1.128)\qquad p(b,t) = \sum_j (y(T) - z_d, w_j) w_j(b) \frac{\sin(T-t)\sqrt{\lambda j}}{\sqrt{\lambda j}}.$$

Let us assume that there is no $v \in \mathbf{U}_{\mathrm{ad}}$ such that

$$y(T) = z_d,$$

and let us also assume that b is a strategic point.

There is a number T_0 such that, if $p(b,t) = 0$ on $(\sigma, \sigma + T_0)$, then $(y(T) - z_d, w_j) w_j(b) = 0 \ \forall j$, a situation which is excluded.

Therefore, on every interval of length T_0 contained in $[0,T]$, $T \geq T_0$, there is a set of positive measure where $u = 0$ or $u = M$. (This is an *extremely weak* form of a kind of bang-bang property!)

REMARK 1.3.6. *The above remarks are related to the possible oscillations of solutions of the wave equation. We refer to Haraux* [10], [11], *Cazeneuve and Haraux* [2], *Haraux and Komornik* [14].

1.3.8. Boundary Criterion. Let us consider the state equation (1.83). The *cost function* we consider now is given by

$$(1.129)\qquad J(v) = \int_\Sigma \left[\frac{\partial y(v)}{\partial \nu} - z_d \right]^2 d\Sigma + N \int_0^T v^2 dt.$$

Let us make precise the meaning of (1.129). Given v in $L^2(0,T)$, we know that y is continuous from $[0,T] \to L^2(\Omega)$. Therefore, we have in particular

$$y \in L^2(0,T;L^2(\Omega)).$$

Let $\mathbf{O}$ be an open set contained in Ω and such that $b \notin \mathbf{O}$. Then, if $y_{\mathbf{O}}$ denotes the restriction of y to $\mathbf{O}$, we have that

$$y_{\mathbf{O}} \in L^2(0,T;L^2(\mathbf{O})), \tag{1.130a}$$

$$\Delta y_{\mathbf{O}} = y''_{\mathbf{O}} \in H^{-2}(0,T;L^2(\mathbf{O})); \tag{1.130b}$$

in (1.130b), we use the notation

$$H^{-2}(0,T;L^2(\Omega)) = \text{ dual space of } H_0^2(0,T;L^2(\Omega)),$$

where

$$H_0^2(0,T;L^2(\Omega)) \\ = \{\phi|\phi,\phi',\phi'' \in L^2(0,T;L^2(\Omega)), \phi(0)=\phi'(0)=\phi(T)=\phi'(T)=0\}.$$

It follows from (1.30b) and the fact that

$$y_{\mathbf{O}}\,|_{\Sigma} = \text{trace of } y_{\mathbf{O}} \text{ on } \Sigma = 0$$

that

$$y_{\mathbf{O}} \in H^{-2}(0,T;H^2(\mathbf{O}_1)), \qquad \mathbf{O}_1 \subset \mathbf{O}.$$

But since $\mathbf{O}$ is arbitrary (with $b \notin \mathbf{O}$), we see that

$$y_{\mathbf{O}} \in H^{-2}(0,T;H^2(\mathbf{O})) \tag{1.131}$$

so that finally

$$\frac{\partial}{\partial \nu} y_{\mathbf{O}} \in H^{-2}(0,T;H^{1/2}(\Gamma)). \tag{1.132}$$

Therefore $\frac{\partial y(v)}{\partial \nu}$ makes sense, and we can define the new function space $\mathbf{U}$:

$$\mathbf{U} = \left\{ v|v \in L^2(0,T), \frac{\partial y(v)}{\partial \nu} \in L^2(\Sigma) \right\}. \tag{1.133}$$

Provided with the graph norm

$$\|v\|_{\mathbf{U}} = \left(\|v\|^2_{L^2(0,T)} + \left\| \frac{\partial y(v)}{\partial \nu} \right\|^2_{L^2(\Sigma)} \right)^{1/2} \tag{1.134}$$

$\mathbf{U}$ becomes a Hilbert space.

Let $\mathbf{U}_{\mathbf{ad}}$ be a closed convex subset of $\mathbf{U}$. We are looking for

(1.135) $$\inf J(v), \quad v \in \mathbf{U}_{\mathrm{ad}}.$$

Problem (1.135) admits a unique solution u. We want to characterize u by the O.S.

In order to do so, let us give some properties of the space $\mathbf{U}$.

1.3.9. Properties of the Space U.

(1.136) In a neighborhood of T, $\mathbf{U}$ coincides with L^2 functions.

Indeed changing v in such a neighborhood (whose size depends on b) does not change $\partial y(v)/\partial\nu$ on Σ, hence (1.136); this remark also implies that $\mathbf{U}$ depends on b.

We verify now that

(1.137) $$\mathbf{D}\ (]0,T[) \text{ is dense in } \mathbf{U}.$$

Indeed, we can, according to the above remark, cut off v in the neighborhood of T without affecting $\partial y(v)/\partial\nu$. Therefore functions of $\mathbf{U}$ which are identically equal to zero near T are dense in $\mathbf{U}$.

We then consider $v_h(t) = v(t-h)$ if $t \geq h$, 0 if $t < h$, with h small enough. The solution y_h corresponding to v_h is simply $y_h(t) = y(t-h)$ for $t > h$, 0 for $t < h$, where $y(v) = y$. Therefore $\partial y_h/\partial\nu \to \partial y/\partial\nu$ in $L^2(\Sigma)$ as $h \to 0$, so that $v_h \to v$ in $\mathbf{U}$. Therefore functions of $\mathbf{U}$ with compact support in $]0,T[$ are dense in $\mathbf{U}$. We can next regularize these functions by convolution in t; hence (1.137) follows.

The optimality system is now given as follows: the optimal control u, solution of (1.135), is characterized by the solution $\{u, y, p\}$ of

(1.138a) $$y'' - \Delta y = u\delta(x-b)$$

(1.138b) $$p'' - \Delta p = 0,$$

(1.138c) $$y = 0, \quad p = -\left(\frac{\partial y}{\partial\nu} - z_d\right) \quad \text{on } \Sigma,$$

(1.138d) $$y(0) = y'(0) = 0, \quad p(T) = p'(T) = 0 \quad \text{in } \Omega,$$

and

(1.139) $$\int_0^T (p(b,t) + Nu(t))(v-u)dt \geq 0 \quad \forall v \in \mathbf{U}_{\mathrm{ad}}.$$

REMARK 1.3.7. *The adjoint state p, defined by* (1.138b) *and*

$$p = -\left(\frac{\partial y}{\partial \nu} - z_d\right) \in L^2(\Sigma), \qquad p(T) = p'(T) = 0,$$

satisfies

$$p(b,t) \in \mathbf{U}', \tag{1.140}$$

where $\mathbf{U} \subset L^2(0,T) \subset \mathbf{U}' \subset \mathbf{D}'(]0,T[)$.

This follows from the definition of $\mathbf{U}$ and Green's formula.

In (1.139), $\int_0^t p(b,t)(v-u)dt$ expresses the duality between $\mathbf{U}'$ and $\mathbf{U}$.

1.3.10. Duality. Let us return to problem (1.86). If we set

$$\begin{aligned} F(v) &= N\int_0^T v^2 dt, \qquad v \in L^2(0,T), \\ G'(q) &= \int_\Omega (q - z_d)^2 dx, \qquad q \in L^2(\Omega), \\ Lv &= y(T;v), \qquad L \in \mathbf{L}(L^2(0,T); L^2(\Omega)), \end{aligned} \tag{1.141}$$

we can apply duality theory in the sense of §1.2.7. This leads to the following class of problems.

The *state equation* is now given by

$$\begin{aligned} &y'' - \Delta y = 0 \quad \text{in } Q, \\ &y(0) = 0, \qquad y'(0) = v \quad \text{in } \Omega, \\ &y = 0 \quad \text{on } \Sigma; \end{aligned} \tag{1.142}$$

let $y(x,t;v) = y(v)$ be the solution of (1.42). Then (cf. §1.3.4)

$$y(b,t;v) \in L^2(0,T),$$

and we can define

$$J(v) = \int_0^T |y(b,t;v) - z_d|^2\, dt + N\int_\Omega v^2 dx. \tag{1.143}$$

Let us consider

$$\mathbf{U}_{\text{ad}} = \text{ closed convex set in } L^2(\Omega). \tag{1.144}$$

We want to find

$$\inf J(v), \quad v \in \mathbf{U}_{\text{ad}}. \tag{1.145}$$

This problem admits a unique solution u, which is characterized by the solution $\{u,y,p\}$ of the O.S.

$$
\begin{aligned}
&y'' - y = 0,\\
&p'' - \Delta p = (y(b,t) - z_d)\delta(x-b) \quad \text{in } Q,\\
&y(0) = 0, \qquad y'(0) = u,\\
&p(T) = p'(T) = 0 \quad \text{in } \Omega,\\
&y = p = 0 \quad \text{on } \Sigma
\end{aligned}
\tag{1.146}
$$

and

$$
\int_\Omega (p(x,0) + Nu(x))(v(x) - u(x))dx \geq 0 \quad \forall v \in \mathbf{U}_{\mathrm{ad}}. \tag{1.147}
$$

1.3.11. Anti-sound. Let us make here a very simple remark concerning the "anti-sound" problem, mentioned in the Introduction (cf. Banks, Propst, and Silcox [1]). The *state* is given by

$$
y'' - \Delta y = g(t)\delta(x-c) + v(t)\delta(x-b) \tag{1.148}
$$

where b and c are given in Ω, $b \neq c$, $g(t)$ given in $L^2(0,T)$ and where v is, as before, the control variable. Let us assume (for simplicity) that

$$
\begin{aligned}
&y(0) = y'(0) = 0 \quad \text{in } \Omega,\\
&y = 0 \quad \text{on } \Sigma.
\end{aligned}
\tag{1.149}
$$

Let ω be an open set given in Ω,

$$
\omega \subset \Omega, \quad b \notin \omega, \quad c \notin \omega. \tag{1.150}
$$

If $y(x,t;v) = y(v)$ denotes the state of the system, solution of (1.148) (1.149), the *cost function* is given by

$$
\begin{aligned}
J(v) &= \int_{\omega x(0,T)} (y(v) - z_d)^2\, dx\, dt + N \int_0^T v^2 dt,\\
&\quad z_d \in L^2(\omega \times (0,T)), \quad N > 0.
\end{aligned}
\tag{1.151}
$$

Let us take

$$
\mathbf{U}_{\mathrm{ad}} = \text{ closed convex subset of } L^2(0,T). \tag{1.152}
$$

Then the problem

$$
\inf J(v), \quad v \in \mathbf{U}_{\mathrm{ad}} \tag{1.153}
$$

admits a unique solution. This solution is characterized by the O.S. given by

$$
(1.154) \qquad \begin{aligned}
&y'' - \Delta y = g\delta(x-c) + u\delta(x-b) \\
&p'' - \Delta p = (y - z_d)\chi_\omega, \qquad \chi_\omega = \text{ characteristic function of } \omega \\
&y(0) = y'(0) = 0, \quad p(T) = p'(T) = 0 \quad \text{in } \Omega, \\
&y = p = 0 \quad \text{on } \Sigma,
\end{aligned}
$$

and

$$
(1.155) \qquad \int_0^T (p(b,t) + Nu)(v-u)dt \geq 0 \quad \forall v \in \mathbf{U}_{\mathbf{ad}}.
$$

REMARK 1.3.8. *Let us emphasize again that this is a very simple preliminary remark on the anti-sound problem! It would be interesting to study—by the methods of A. Haraux* [10], [11]*—the possible oscillations of* $p(b,t)$*, in order (in particular) to study the bang-bang properties of* u.

REMARK 1.3.9. *Let us set*

$$
(1.156) \qquad \mathbf{Y}(b) = \inf J(v), \qquad v \in \mathbf{U}_{\mathbf{ad}}.
$$

In the "anti-sound" setting, it is particularly natural to study the function

$$
(1.157) \qquad b \to \mathbf{Y}(b), \qquad \Omega \to \mathbb{R}.
$$

This is an example of a very general question. This type of question has been studied in Lions [21, Chap. 4, §4], for diffusion equations, when $\mathbf{U}_{\mathbf{ad}} = \mathbf{U}$.

If we consider the case

$$
(1.158) \qquad \mathbf{U}_{\mathbf{ad}} = L^2(0,T),
$$

then $p(b,t) + Nu = 0$, so that (1.154), (1.155) reduces to

$$
(1.159) \qquad \begin{aligned}
&y'' - \Delta y + \frac{1}{N} p\delta(x-b) = g\delta(x-c), \\
&p'' - \Delta p = (y - z_d)\chi_\omega, \\
&y(0) = y'(0) = 0, \quad p(T) = p'(T) = 0 \quad \text{in } \Omega, \\
&y = p = 0 \quad \text{on } \Sigma.
\end{aligned}
$$

We have

$$
(1.160) \qquad \mathbf{Y}(b) = -\int_{\omega x(0,T)} (y - z_d) z_d \, dx \, dt + \int_0^T p(c,t) g(t) \, dt.
$$

Let us compute, formally for the time being, $(\partial/\partial b_i)\mathbf{Y}(b)$. If we set

$$
\dot{y} = \frac{\partial y}{\partial b_i}, \qquad \dot{p} = \frac{\partial p}{\partial b_i},
$$

we have

$$\frac{\partial}{\partial b_i}\mathbf{Y}(b) = -\int_{\omega x(0,T)} \dot{y} z_d dx\, dt + \int_0^T \dot{p}(c,t)g(t)dt. \tag{1.161}$$

This formula makes sense only if z_d and g satisfy some regularity properties. The couple $\{\dot{y}, \dot{p}\}$ satisfies

$$\begin{aligned} &\dot{y}'' - \Delta\dot{y} + \frac{1}{N}\dot{p}\delta(x-b) + \frac{1}{N}p\frac{\partial\delta}{\partial x_i}(x-b) = 0, \\ &\dot{p}'' - \Delta\dot{p} = \dot{y}\chi_\omega, \\ &\dot{y}(0) = \dot{y}'(0) = 0, \quad \dot{p}(T) = \dot{p}'(T) = 0 \quad \text{in } \Omega, \\ &\dot{y} = \dot{p} = 0 \quad \text{on } \Sigma. \end{aligned} \tag{1.162}$$

Due to the presence of $(\partial/\partial x_i)\delta(x-b)$ in (1.162), one has

$$\dot{y} \text{ is continuous from } [0,T] \to H^{-1}(\Omega). \tag{1.163}$$

One has the same property for $\dot{p}$; outside ω, one has

$$\Delta\dot{p} = \dot{p}'' \in H^{-2}(0,T;H^{-1}(\mathbf{O})), \qquad \mathbf{O} \subset \Omega, \quad \mathbf{O} \cap \omega = \phi$$

so that

$$\dot{p} \in H^{-2}(0,T;H^1(\mathbf{O}_1)), \quad \text{where } \mathbf{O}_1 \subset \mathbf{O}, \tag{1.164}$$

then

$$\Delta\dot{p} = \dot{p}'' \in H^{-4}(0,T;H^1(\mathbf{O}_1))$$

and

$$\dot{p} \in H^{-4}(0,T;H^3(\mathbf{O}_2)), \qquad \bar{\mathbf{O}}_2 \subset \mathbf{O}. \tag{1.165}$$

By interpolation between (1.164) and (1.165), one has

$$\dot{p} \in H^{-\alpha}(0,T;H^\beta(\mathbf{O}_2)), \qquad \alpha = 1+\beta. \tag{1.166}$$

We choose β such that $H^\beta(\mathbf{O}_2) \subset$ continuous functions in $\mathbf{O}_2$, i.e., $\beta > n/2$. Therefore

$$\dot{p}(c,t) \in H^{-(1+\frac{n}{2}+\varepsilon)}(0,T), \qquad \varepsilon > 0. \tag{1.167}$$

It follows from (1.163), (1.167) that (1.161) makes sense—and actually this can be justified—if

$$\begin{aligned} &z_d \in L^2(0,T;H_0^1(\omega)), \\ &g \in H_0^{1+\frac{n}{2}+\varepsilon}(0,T) \quad \text{for some } \varepsilon > 0. \end{aligned} \tag{1.168}$$

REMARK 1.3.10. *An interesting situation arises when $g(t)$ is only partly known in* (1.148). *If we assume that*

$$g \in G, \qquad G = \text{ closed subspace of } L^2(0,T) \tag{1.169}$$

we think of the solution of (1.148), (1.149) *as a function of v and of g:*

$$y = y(v,g). \tag{1.170}$$

Then $J(v)$ becomes

$$J(v,g) = \int_{\omega x(0,T)} [y(v,g) - z_d]^2 \, dx \, dt + N \int_0^T v^2 \, dt \tag{1.171}$$

and one can look for Pareto controls *associated with this problem. We refer to Lions* [25], [26] *for the first notions concerning Pareto controls for distributed systems.*

1.3.12. Pointwise Control on the Boundary. Let us now consider situations where one can act on the system in a pointwise manner and only on the boundary of the domain.

The state of the system is given by

$$\begin{aligned} &y'' - \Delta y = 0 \quad \text{in } Q = \Omega \times]0,T[, \\ &y = v(t)\delta_\Gamma(b) \quad \text{on } \Sigma \\ &y(0) = y'(0) = 0 \quad \text{in } \Omega; \end{aligned} \tag{1.172}$$

in (1.172) $\delta_\Gamma(b)$ denotes the Dirac mass on Γ at point $b \in \Gamma$, i.e.,

$$\int_\Gamma \delta_\Gamma(b)\phi \, d\Gamma = \phi(b) \quad \forall \phi \in C^0(\Gamma). \tag{1.173}$$

REMARK 1.3.11. *Another problem would be obtained by replacing in* (1.172) *the nonhomogeneous Dirichlet condition by a nonhomogeneous Neumann condition, namely*

$$\frac{\partial y}{\partial \nu} = v(t)\delta_\Gamma(b). \tag{1.174}$$

REMARK 1.3.12. *Of course, one deals here with weak solutions. If ϕ is the solution of*

$$\phi'' - \Delta\phi = \psi \quad in\, Q, \quad \phi(T) = \phi'(T) = 0, \quad \phi = 0 \quad on\, \Sigma, \tag{1.175}$$

then, if y is (formally) a solution of (1.172), *we obtain*

$$\int_Q y\psi \, dx \, dt = -\int_0^T v(t)\frac{\partial \phi}{\partial \nu}(b,t) \, dt. \tag{1.176}$$

For ψ smooth enough,

$$\psi \to -\int_0^T v(t)\frac{\partial \phi}{\partial \nu}(b,t)\,dt$$

is defined—assuming that, say, $v \in L^2(0,T)$*—and* $y = y(v)$ *is defined as a weak solution of* (1.172).

A similar construction will apply to case (1.174). *This time, one takes* $\partial\phi/\partial\nu = 0$ *on* Σ *in* (1.175) *and obtains*

$$\int_Q y\Psi\,dx\,dt = \int_0^T v(t)\phi(b,t)\,dt. \tag{1.177}$$

REMARK 1.3.13. *We conjecture that, if* $v \in L^2(0,T)$ *then the solution* $y(v)$ *of* (1.172) *satisfies (when* $\Omega \subset \mathbb{R}^3$*)*

$$t \to y(t;v) \text{ is continuous from } [0,T] \to H^{-1}(\Omega). \tag{1.178}$$

The dual property *of* (1.178) *is the following: we consider*

$$\begin{aligned} &\phi'' - \Delta\phi = 0 \quad \text{in } \Omega\times]0,T[,\\ &\phi(0) = 0, \qquad \phi'(0) = h \quad \text{in } \Omega,\\ &\phi = 0 \quad \text{on } \Sigma; \end{aligned} \tag{1.179}$$

then an equivalent conjecture is

$$\int_0^T \left|\frac{\partial\phi}{\partial\nu}(b,t)\right|^2 dt \le c\|h\|^2_{H_0^1(\Omega)}. \tag{1.180}$$

Inequality (1.180) *is true if* $\Omega = \mathbb{R}^3$*; one has just to apply* (1.119) *after taking* $\partial/\partial x_i$ *derivatives of the equations. It is also true if* $\Omega =$ *half space; by a simple reflexion argument, one reduces the problem to the case* $\Omega = \mathbb{R}^3$.

If one introduces, in the general case, $\tilde{h} =$ *extension of* h *to* $\mathbb{R}^3$ *by zero outside* Ω*, and one defines* ϕ *by*

$$\begin{aligned} &\phi'' - \Delta\phi = 0 \quad \text{in } \mathbb{R}^3 \times \{t < 0\},\\ &\phi(0) = 0, \qquad \phi'(0) = h, \end{aligned} \tag{1.181}$$

then

$$w = \phi_\Omega - \phi \qquad (\phi_\Omega = \text{ restriction of } \phi \text{ to } \Omega) \tag{1.182}$$

satisfies

$$\begin{aligned} &w'' - \Delta w = 0,\\ &w(0) = w'(0) \quad \text{in } \Omega,\\ &w = \phi \quad \text{on } \Sigma. \end{aligned} \tag{1.183}$$

Using then the result of Lasiecka, Lions, and Triggiani [19] *one gets "close" to prove* (1.180)*, without obtaining the desired result, which remains open.*

REMARK 1.3.14. *We do not offer any conjecture on the best possible regularity result for the case* (1.174). *This is related to regularity problems for the wave equation with Neumann boundary conditions; cf. Lasiecka and Triggiani* [17].

If we admit (1.178), a natural optimal control problem for (1.172) would be

$$J(v) = \|y(T;v) - z_d\|^2_{H^{-1}(\Omega)} + N\int_0^T v^2\,dt. \tag{1.184}$$

In (1.184), z_d is given in $H^{-1}(\Omega)$, N is given > 0; the norm in $H^{-1}(\Omega)$ is defined as follows:

$$\|f\|^2_{H^{-1}(\Omega)} = (Gf, f) \tag{1.185}$$

where

$$-\Delta Gf = f \quad \text{in } \Omega, \qquad Gf = 0 \quad \text{on } \Gamma, \tag{1.186}$$

and where in (1.185) (Gf, f) expresses the duality between $H_0^1(\Omega)$ and $H^{-1}(\Omega)$.

If

$$\mathbf{U}_{\mathrm{ad}} = \text{ closed convex set in } L^2(0,T) \tag{1.187}$$

the problem is to find

$$\inf J(v), \quad v \in \mathbf{U}_{\mathrm{ad}}, \tag{1.188}$$

which admits a unique solution u. The *optimal control* u is characterized by the O.S.

$$\begin{aligned} &y'' - \Delta y = 0,\\ &p'' - \Delta p = 0,\\ &y = u\delta_\Gamma, \quad p = 0 \quad \text{on } \Sigma,\\ &y(0) = y'(0) = 0,\\ &p(T) = 0, \quad p'(T) = G(y(T) - z_d) \end{aligned} \tag{1.189}$$

and

$$\int_0^T \left(\frac{\partial p}{\partial \nu}(b,t) + Nu(t)\right)(v(t) - u(t))dt \geq 0 \quad \forall v \in \mathbf{U}_{\mathrm{ad}}. \tag{1.190}$$

(In (1.190) we use the "equivalent conjecture" (1.180).)

1.3.13. Some Extensions.

REMARK 1.3.15. *We can consider variable points b, i.e.,* $b = b(t)$.

Let us assume that

$$(1.191)\qquad \begin{aligned} &t \to b(t) \text{ is a } C^1 \text{ mapping from } [0,T] \to \Omega,\\ &b(t) \in K = \text{ compact subset of } \Omega. \end{aligned}$$

Let the state equation be given by

$$(1.192)\qquad \begin{aligned} &y'' - \Delta y = v(t)\delta(x - b(t)) \quad \text{in } Q\\ &y(x,0) = y'(x,0) = 0 \quad \text{in } \Omega,\\ &y = 0 \quad \text{on } \Sigma. \end{aligned}$$

This problem admits a unique solution $y(v)$. One can show—by methods entirely similar to those of §§1.3.3 and 1.3.4—that, if $v \in L^2(0,T)$, then

$$(1.193)\qquad t \to y(t;v) \text{ is continuous from } [0,T] \to L^2(\Omega).$$

Let the cost function be given by

$$(1.194)\qquad J(v) = \int_\Omega [y(x,T;v) - z_d]^2\,dx + N\int_0^T v^2\,dt.$$

If $\mathbf{U}_{\text{ad}}$ equals a closed convex subset of $L^2(0,T)$, then

$$(1.195)\qquad \inf J(v), \quad v \in \mathbf{U}_{\text{ad}}$$

admits a unique solution u which is characterized by the *O.S.*

$$(1.196)\qquad \begin{aligned} &y'' - \Delta y = u\delta(x - b(t)),\\ &p'' - \Delta p = 0,\\ &y(0) = y'(0) = 0, \quad p(T) = 0, \quad p'(T) = -(y(T) - z_d) \quad \text{on } \Omega,\\ &y = p = 0 \quad \text{on } \Sigma, \end{aligned}$$

and

$$(1.197)\qquad \int_0^T (p(b(t),t) + Nu(t))(v(t) - u(t))\,dt \geq 0 \quad \forall v \in \mathbf{U}_{\text{ad}}.$$

REMARK 1.3.16. *The analysis of the situation where the distance from* $b(t)$ *to* Γ *can tend to zero is open.*

REMARK 1.3.17. *All that has been said extends to the case when* $-\Delta$ *is replaced by a second-order elliptic operator, as in Remark* 1.3.1. *It also extends to hyperbolic systems, but technical details are not presented here.*

1.4. Some Other Topics.

1.4.1. Controllability. Let us consider now the following question: let the state of the system be given by

$$y'' - \Delta y = v(t)\delta(x - b), \quad b \text{ given in } \Omega, \tag{1.198}$$

with

$$y(0) = y^0, \quad y'(0) = y^1 \quad \text{in } \Omega \tag{1.199}$$

$$y = 0 \quad \text{on } \Sigma. \tag{1.200}$$

In (1.199) y^0, y^1 are given (in suitable function spaces to be defined later) and we assume that

$$v \in L^2(0, T). \tag{1.201}$$

The question is the following (problem of exact controllability): given T, is it possible to find v such that

$$y(T; v) = 0, \qquad y'(T; v) = 0. \tag{1.202}$$

REMARK 1.4.1. *Due to the finite speed of propagation, in order to find v such that* (1.202) *holds true, it is necessary for T to be large enough. For instance, if $T \leq d(b, \Gamma)$ = distance from b to Γ, we cannot change the behavior of the solution in the region of Ω outside the ball centered at b and with radius $d(b, \Gamma)$.*

We initiate here the solution of the exact controllability problem by using the systematic method indicated in Lions [27], [28], [29].

Let ϕ^0, ϕ^1 be given, with

$$\phi^0 \in H_0^1(\Omega), \qquad \phi^1 \in L^2(\Omega) \tag{1.203}$$

and let us define ϕ as the solution of

$$\begin{aligned} &\phi'' - \Delta\phi = 0, \\ &\phi(0) = \phi^0, \qquad \phi'(0) = \phi^1 \quad \text{in } \Omega, \\ &\phi = 0 \quad \text{in } \Sigma. \end{aligned} \tag{1.204}$$

Let us set

$$\|\{\phi^0, \phi^1\}\|_F = \left(\int_0^T \phi(b, t)^2 dt\right)^{1/2}. \tag{1.205}$$

Under hypothesis (1.203), $\phi(b,t) \in L^2(0,T)$ (cf. §1.3.5) so that (1.205) makes sense.

Let us assume that b is *strategic* (cf. §1.2.5, Example 1.2.3). Then

$$\phi(b,t) = \sum_j \left[\left(\phi^0, w_j\right) \cos t\sqrt{\lambda_j} + \left(\phi^1, w_j\right) \frac{\sin t\sqrt{\lambda_j}}{\sqrt{\lambda_j}} \right] w_j(b) \tag{1.206}$$

and if $\phi(b,t) = 0$ for $t \in (0,T)$ and if T is large enough, then it seems plausible that $(\phi^0, w_j) w_j(b) = 0$ and $(\phi^1, w_j) w_j(b) = 0$, which implies (since $w_j(b) \neq 0$ for all j) that

$$\left(\phi^0, w_j\right) = 0, \qquad \left(\phi^1, w_j\right) = 0 \quad \forall j,$$

i.e., $\phi^0 = 0, \phi^1 = 0$. Therefore it would seem very plausible (cf. Notes added in proof, p. 37) that

(1.207) if b is strategic and if T is large enough, (1.205) defines a norm on $H_0^1(\Omega) \times L^2(\Omega)$.

Assuming (1.207) to be true, we then define F as the completion of $H_0^1(\Omega) \times L^2(\Omega)$ for the norm (1.205), and F' as the dual of F. We now define an operator $\Lambda \in \mathbf{L}(F; F')$ as follows: let ψ be the solution of

$$\begin{aligned} &\psi'' - \Delta\psi = -\phi(b,t)\delta(x-b) \quad \text{in } Q, \\ &\psi(T) = \psi'(T) = 0 \quad \text{in } \Omega, \\ &\psi = 0 \quad \text{on } \Sigma. \end{aligned} \tag{1.208}$$

Since $\phi(b,t) \in L^2(0,T), \psi$ is continuous from $[0,T] \to L^2(\Omega)$ and ψ' is continuous from $[0,T] \to H^{-1}(\Omega)$, so that

$$\psi(0) \in L^2(\Omega), \qquad \psi'(0) \in H^{-1}(\Omega), \tag{1.209}$$

and we can define

$$\Lambda\{\phi^0, \phi^1\} = \{\psi'(0), -\psi(0)\}. \tag{1.210}$$

Let us show that

$$\Lambda \in \mathbf{L}(F; F'). \tag{1.211}$$

In a general way, let us define z as the solution of

$$z'' - \Delta z = g(t)\delta(x-b) \quad \text{in } Q, \quad g \in L^2(0,T), \tag{1.212a}$$

$$z(T) = z'(T) = 0 \quad \text{in } \Omega, \tag{1.212b}$$

$$z = 0 \quad \text{on } \Sigma. \tag{1.212c}$$

Then, if we multiply (1.212a) by ϕ and if we integrate by parts (assuming all data to be smooth), we obtain

$$\int_Q \phi(b,t)g(t)dt = -(z'(0),\phi(0)) + (z(0),\phi'(0)). \tag{1.213}$$

Therefore, if $g \in L^2(0,T)$,

$$\{\phi^0,\phi^1\} \to (z'(0),\phi^0) - (z(0),\phi^1)$$

is a continuous linear form on F, so that

$$\{z'(0),-z(0)\} \in F'.$$

Applying this remark with $g(t) = -\phi(b,t)$, we find (1.211) and moreover, according to (1.213):

$$\left\langle \Lambda\{\phi^0,\phi^1\},\{\phi^0,\phi^1\}\right\rangle = \left\|\{\phi^0,\phi^1\}\right\|_F^2 \tag{1.214}$$

where the bracket expresses the duality between F' and F.

By virtue of (1.214) Λ is an isomorphism from F onto F'. Therefore, given y^0, y^1 such that $\{y^1,y^0\} \in F'$, there exists a unique $\{\phi^0,\phi^1\} \in F$ such that

$$\Lambda\{\phi^0,\phi^1\} = \{y^1,-y^0\}. \tag{1.215}$$

If we choose now

$$\begin{aligned} &v = -\phi(b,t),\\ &\phi \text{ solution of (1.204) with } \{\phi^0,\phi^1\} \text{ given by (1.215)} \end{aligned} \tag{1.216}$$

then the corresponding state $y(t;v)$ satisfies (1.202) (since y coincides then with ψ solution of (1.208), since $\{\psi'(0),-\psi(0)\} = \{y^1,-y^0\}$).

This shows exact controllability under conditions (1.207) and assuming that $\{y^1,y^0\} \in F'$. But (1.207) may be false, as shown by Y. Meyer and S. Jaffard.

REMARK 1.4.2. *We have*

$$H_0^1(\Omega) \times L^2(\Omega) \subset F, \qquad F' \subset H^{-1}(\Omega) \times L^2(\Omega).$$

It would be interesting to obtain more precise information on F (cf. Haraux [12], Lions [29]).

REMARK 1.4.3. *We have seen (cf. (1.207)) that if $T > T_0$, (1.205) defines a norm. Let T be given.*

It seems extremely likely that for T' such that

$$T_0 < T' < T$$

the norm defined by $(\int_0^{T'} \phi(b,t)^2\,dt)^{1/2}$ is equivalent to the norm given by (1.205), so that we define the same space F when using $(\int_0^{T'} \phi(b,t)^2\,dt)^{1/2}$.

Given then y^0, y^1 such that $\{y^1, y^0\} \in F'$, there exists an *infinite* number of controls v such that

$$y(T; v) = y'(T; v) = 0.$$

We then look for

$$\inf_{v \in L^2(0,T)} \int_0^T v^2 dt, \quad v \text{ subject to } (1.202). \tag{1.217}$$

Let u be the (unique) solution of (1.217). This optimal u is the control given by (1.216). Cf. Lions [29] for details.

REMARK 1.4.4. *The method indicated above is completely general.*

We refer to Lions [28], [29] for many other examples. For Dirichlet's type boundary control, cf. Lop Fat Ho [15], [16]. For numerical algorithms corresponding to the general method, we refer to R. Glowinski, C. Li, and J. L. Lions [9]; cf. also [4].

REMARK 1.4.5. *If the eigenvalues are, say, at most of multiplicity* 2, *then two (suitable) points* b^1, b^2 *in* Ω *will be needed for exact controllability, the state equation being given by*

$$y'' - \Delta y = v_1(t)\delta(x - b^1) + v_2(t)\delta(x - b^2). \tag{1.218}$$

1.4.2. Stabilizability. It is well known (cf. Russell [39], [40]) that exact controllability implies the existence of *feedback* implying exponential decrease of the solution.

Let us mention here a simple direct approach, leading to an interesting open question.

Let us consider, for $t > 0$, the equation

$$y'' - \Delta y = v(t)\delta(x - b(t)), \tag{1.219a}$$

$$y(0) = y^0, \qquad y'(0) = y^1 \tag{1.219b}$$

$$y = 0 \quad \text{on } \Sigma, \tag{1.219c}$$

where $y^0 \in L^2(\Omega), y^1 \in H^{-1}(\Omega)$. We want to choose (if possible!) the function v in (1.219b) so as to insure "fast decrease" of y.

If we start with

$$\{y^0, y^1\} \in L^2(\Omega) \times H^{-1}(\Omega) \tag{1.220}$$

it is natural to consider the "energy" given by

$$\frac{1}{2}[|y(t)|^2 + (Gy', y')] = E(t), \tag{1.221}$$

where $|y(t)|^2 = \int_\Omega y(t)^2\,dx$ and where $G = (-\Delta)^{-1}$ for Dirichlet boundary condition on Γ (so that $(Gy', y') = \|y'\|^2_{H^{-1}(\Omega)}$). Then

$$\frac{d}{dt}E(t) = (y', Gy'' + y). \tag{1.222}$$

But applying G to (1.219a) gives

$$Gy'' + y = vG\delta(x - b(t)) = v(t)G(x, b(t))$$

if $G(x,\xi) =$ kernel of G (Green's kernel). Therefore (1.222) gives

$$\frac{d}{dt}E(t) = v(t)\int_\Omega y'(x,t)G(x,b(t))\,dx. \tag{1.223}$$

An "obvious" candidate for $v(t)$ is then

$$v(t) = -\int_\Omega y'(x,t)G(x,b(t))\,dx, \tag{1.224}$$

hence, we obtain the equation

$$\begin{aligned} &y'' - \Delta y + \left(\int_\Omega y'(x,t)G(x,b(t))\,dx\right)\delta(x - b(t)) = 0,\\ &y(0) = y^0, \qquad y'(0) = y^1,\\ &y = 0 \quad \text{on } \Sigma. \end{aligned} \tag{1.225}$$

It would be very interesting to know the behavior as $t \to +\infty$ of the solution of (1.225), at least when

$$b(t) = b = \text{ strategic point.}$$

REMARK 1.4.6. *For boundary control, somewhat similar feedback has been proposed in Lasiecka and Triggiani* [18], *where the authors prove an* exponential decrease *in time of the solution, under some geometrical conditions on* Ω.

1.4.3. Petrowski Equations. In all that has been said in §1.3, the hyperbolic character of the model does not always play an essential role.

We can, for instance, consider the Petrowsky type of equation

$$\begin{aligned} &y'' + \Delta^2 y = v(t)\delta(x - b),\\ &y = \frac{\partial y}{\partial \nu} = 0 \quad \text{on } \Sigma,\\ &y(x,0) = y'(x,0) = 0 \quad \text{in } \Omega. \end{aligned} \tag{1.226}$$

If we consider the *cost function*

$$J(v) = \int_\Omega [y(x,T;v) - z_d(x)]^2 dx + N\int_0^T v^2\,dt, \tag{1.227}$$

then there is a unique optimal control u solution of

(1.228) $$J(u) = \inf J(v),$$

(1.229) $$v \in \mathbf{U}_{\mathrm{ad}} = \text{ closed convex subset of } L^2(0,T).$$

REMARK 1.4.7. *It is a simple matter to verify that if* $v \in L^2(0,T), \Omega \subset \mathbb{R}^3$, *then* y *is continuous from* $[0,T] \to L^2(\Omega)$.

Much more difficult seems to be the (likely) property that y is continuous from $[0,T] \to H^{3/2}(\Omega)$ (cf. Remark 1.3.4, §1.3.5).

The O.S., which characterizes the optimal control u is given by

(1.230)
$$\begin{aligned}
&y'' + \Delta^2 y = u\delta(x-b),\\
&p'' + \Delta^2 p = 0 \quad \text{in } Q,\\
&y(0) = y'(0) = 0, \quad p(T) = 0, \quad p'(T) = -(y(T) - z_d) \quad \text{in } \Omega,\\
&y = \frac{\partial y}{\partial \nu} = 0, \quad p = \frac{\partial p}{\partial \nu} = 0 \quad \text{on } \Sigma,
\end{aligned}$$

with

(1.231) $$\int_0^T (p(b,t) + Nu)(v-u)dt \geq 0 \quad \forall v \in \mathbf{U}_{\mathrm{ad}}.$$

1.4.4. Perturbations. Let us consider now the state equation given by

(1.232)
$$\begin{aligned}
&y_\varepsilon'' + \varepsilon\Delta^2 y_\varepsilon - \Delta y_\varepsilon = v(t)\delta(x-b) \quad \text{in } Q,\\
&y_\varepsilon(0) = y_\varepsilon'(0) = 0 \quad \text{in } \Omega,\\
&y_\varepsilon = \frac{\partial y_\varepsilon}{\partial \nu} = 0 \quad \text{on } \Sigma.
\end{aligned}$$

Let $y_\varepsilon(v)$ be the solution of (1.232); $t \to y_\varepsilon(t;v)$ is continuous from $[0,T] \to L^2(\Omega)$. The *cost function* is given now by

(1.233) $$J_\varepsilon(v) = \int_\Omega [y_\varepsilon(x,T;v) - z_d(x)]^2\,dx + N\int_0^T v^2\,dt.$$

Let $\mathbf{U}_{\mathrm{ad}}$ be given as in (1.229) and let u be the solution of

(1.234) $$J_\varepsilon(u_\varepsilon) = \inf J_\varepsilon(v), \qquad v \in \mathbf{U}_{\mathrm{ad}}.$$

It is now a simple matter to verify that, as $\varepsilon \to 0$,

(1.235) $$u_\varepsilon \to u \quad \text{in } L^2(0,T)$$

where u is the solution of (1.86).

REMARK 1.4.8. *Much more complex situations are studied in Lions* [30].

Notes Added in Proof.

(1) The comments following (1.207) in the text have been modified from the first version so as to take into account the (amazing, in the author's opinion) counterexamples found by Y. Meyer and S. Jaffard.

(2) The exact controllability problems alluded to in the text have been the subject of a rather large number of publications. A general presentation of H.U.M. is given in Lions [27].

Further developments have appeared since—for the wave equation very precise results have been obtained by C. Bardos, G. Lebeau, and J. Rauch, ["Sharp sufficient conditions for the observation, control and stabilization of waves from the boundary," *SIAM J. Control*, 30 (1992), pp. 1024–1065]. Even more systematic and general results have been obtained, by D. Tataru, to appear, which introduces interesting arguments.

For the plates equations, cf. J. Lagnese and J. L. Lions, *Modelling, Analysis and Control of Thin Plates*, Masson, R.M.A., 1988, and the bibliography therein. Further developments have since been given by several authors. For Schrödinger and for plates equations, we indicate the works of A. Bensoussan, V. Komornik, G. Lebeau, and E. Zuazua where HUM is used in several different ways. For Maxwell's equation, we refer to J. Lagnese and to O. Nalin (C. R. Acad. Sci. Paris, 1989), all these works are in publication. Several open questions remain for pointwise control and also for pointwise stabilization.

This last question of stabilization seems very interesting also in nonlinear equations. Can we find an action on a "small" subset of the domain Ω where the equation is considered which would "stabilize" the system, such as making the solution exponentially decaying? Results along these lines have been obtained by E. Zuazua, to appear.

(3) One can also consider situations as in the present chapter *and* as in [3] (with variable b's).

References

[1] H. T. Banks, G. Propst, and R. Silcox, *Distributed Systems Control Problems Arising in Active Noise Suppression*, 1986, preprint.

[2] T. Cazenave and A. Haraux, *Some oscillatory properties of the wave equation in several spatial dimensions*, to appear.

[3] D. Cioranescu and J. Saint Jean Paulin, *Homogenization in open sets with holes*, J. Math. Pures. Appl., 71 (1979), pp. 590–607.

[4] E. J. Dean and P. Gubernatis, *Pointwise control of Burgers' equation—A numerical approach*, Comp. Math. Appl., 22 (1991), pp. 93–100.

[5] A. El Jaï, *Etude d'algorithmes pour la commande de systèmes à paramètres répartis de type parabolique*, Thèse de Doctorat d'Etat, Université Paul Sabatier, Toulouse, France, 1978.

[6] A. El Jai and W. G. Pritchard, *Capteurs et actionneurs dans l'analyse des systèmes distribués*, R.M.A., 3, Masson, Paris, 1986.

[7] I. Ekeland and R. Temam, *Analyse convexe et problèmes variationnels*, Dunod, Gauthier Villars, Paris, 1974; English translation: North Holland, Amsterdam, 1979.

[8] J. E. Frowls Williams, *Anti-sound*, Proc. Roy. Soc. London, Sect. A, 393 (1984), pp. 63–88.

[9] R. Glowinski, C. Li and J. L. Lions [1] *A numerical approach to the exact boundary controllability of the wave equation* (I). Japanese J. App. Math., (1990), pp. 1–76.

[10] A. Haraux, *Propriétés de solutions de l'équation des ondes avec conditions de Dirichlet au bord*, Séminaire Bony-Sjöstrand-Meyer, (1984–1985), exp. IX.

[11] ——, *Semilinear hyperbolic problems in bounded domains*, Mathematical Reports, Vol. 3, Part I, J. Dieudonné, ed., Harwood Academic, Gordon and Breach, London, 1987.

[12] ——, *Quelques propriétés des séries lacunaires utiles dans l'étude des vibrations élastiques*, Nonlinear Partial Differential Equations and their Applications, Collège de France Seminars, 1988. H. Brézis and J. L. Lions, eds., Pitman Research Notes in Mathematics.

[13] A. Haraux and S. Jaffard, *Pointwise and spectral control of plate vibrations*, Revista Ibero Americana, 7 (1991), pp. 1–24.

[14] A. Haraux and V. Komornik, *Oscillations of an harmonic Fourier series and the wave equation*, Revista Ibero-Americana, 1985.

[15] Lop Fat Ho, *Boundary observability of the wave equation*, C. R. Acad. Sci. Paris Sér. I Math., 302 (1986), pp. 443–446.

[16] ——, *Exact controllability of second order hyperbolic systems with control in the Dirichlet boundary conditions*, J. Math. Pures Appl., 1987.

[17] I. Lasiecka and R. Triggiani, *Exact controllability of the wave equation with Neumann boundary control*, Appl. Math. Optim., 19 (1989), pp. 243–290.

[18] ——, *Uniform exponential energy decay in a bounded region with $L^2(0,\infty;L^2(\Gamma))$ feedback control in the Dirichlet boundary conditions*, J. Differential Equations, 66 1987), pp. 340–390.

[19] I. Lasiecka, J. L. Lions, and R. Triggiani, *Non homogeneous boundary value problems for second order hyperbolic operators*, J. Math. Pures Appl., 65 (1986) pp. 149–192.

[20] J. L. Lions, *Contrôle optimal des systèmes gouvernés par des équations aux dérivées partielles*, Dunod, Gauthier Villars, Paris, 1968; English translation by S. Mitter, Springer-Verlag, Berlin, 1970.

[21] ——, *Some Methods in the Mathematical Analysis of Systems and their Control*; Science Press, Beijing, and Gordon and Breach, New York, 1981.

[22] ——, *Function spaces and optimal control of distributed systems*, Instituto de Matematica, U.F.R.J., Rio de Janeiro, Brazil, 1980.

[23] ——, *Contrôle des systèmes distribués singuliers*, Gauthier Villars, Paris, 1983; English translation, 1985.

[24] ——, *Some remarks on the optimal control of distributed systems*, Ouspechi Mat. Nauk, 40 (1985), pp. 55–68.

[25] ——, *Contrôle de Pareto de systèmes distribués* (I). *Le cas stationnaire*. (II) *Le cas d'évolution*, C. R. Acad. Sci. Paris Sér. I Math., 302 (1986), pp. 223–227, pp. 413–417.

[26] ——, *Pareto Control of Distributed Systems. An Introduction*, Proceedings of IFIP Workshop on Control, Gainesville, FL, February 1986.

[27] ——, *Contrôlabilité exacte*, R.M.A., Vol. 8, Masson, Paris, 1988; *Perturbations*, R.M.A., Vol. 9, Masson, Paris, 1988.

[28] ——, *Contrôlabilité exacte des systèmes distribués*, C. R. Acad. Sci. Paris Ser. I. Math., 302 (1986), pp. 471–475.

[29] ——, *Exact controllability, stabilization and perturbations for distributed systems*, J. von Neumann Lecture, SIAM Annual Meeting, Boston, MA, July 1986, SIAM Rev., 30 (1988), pp. 1–68.

[30] ——, *Exact controllability and singular perturbations, An example*, Seminar dedicated to P. Lax, Berkeley, CA, June 1986, Wave Motion: Theory, Modelling and Computation, A. J. Chorin and A. J. Majda, eds., Springer-Verlag, New York, 1987, pp. 217–267.

[31] J. L. Lions and E. Magenes, *Problèmes aux limites non homogènes et applications*, Dunod, Paris, Vol. 1 and 2 (1968); English translation, Springer-Verlag, Berlin, 1970.

[32] Li Ta Tsien, *Propriétés d'espaces fonctionnels intervenant en contrôle optimal*, C. R. Acad. Sci. Paris, Ser. I. Math., 289 (1979), pp. 687–690.

[33] ——, *Propriétés d'espaces fonctionnels et problèmes de contrôle optimal de systèmes gouvernés par les équations paraboliques*, C. R. Acad. Sci. Paris Ser. I. Math., 290 (1980), pp. 697–700.

[34] Y. Meyer, *Etude d'un modèle mathématique issu du contrôle des structures spatiales déformables*, Nonlinear Partial Differential Equations and Their Applications, Collége de France seminars, 1983–1984, Vol. 7, H. Brezis and J. L. Lions, eds., Res. Notes in Math. 122, Pitman, London 1985, pp. 234–242.

[35] ——, *Exposé séminaire Brezis-Lions*, Pitman, to appear.

[36] ——, *Algebraic Numbers and Harmonic Analysis*, Mathematical Library, Vol. 2, North Holland, Amsterdam, 1972.

[37] L. Nirenberg, Personal communication.

[38] T. R. Rockafellar, *Duality and stability in extremum problems involving convex functions*, Rational J. Math., 21 (1967), pp. 167–187.

[39] D. L. Russell, *A unified boundary controllability theory for hyperbolic and parabolic partial differential equations*, Studies in Appl. Math., 52 (1973), pp. 189–211.

[40] ——, *Controllability and stabilizability theory for linear partial differential equations: recent progress and open questions*, SIAM Rev., 20 (1978), pp. 639–739.

[41] S. C. Shih, *Sur une classe d'espaces fonctionnels intervenant en contrôle optimal*, C. R. Acad. Sci. Paris, 290 (1980), pp. 761–763.

[42] S. C. Shih and J. Simon, *Sur un espace intervenant en contrôle ponctuel*, C. R. Acad. Sci. Paris, 290 (1980), pp. 693–696.

[43] J. Simon, *Caractérisation d'un espace fonctionnel intervenant en contrôle optimal*, Ann. F. Sci. Toulouse, 5 (1983), pp. 149–169.

Chapter 2

On Issues Related to Stabilization of Large Flexible Structures

M. C. Delfour and M. P. Polis

Foreword.

This paper was written in October 1986 and revised in final form in August 1987. So it is more than two years old. It represents the situation at that time. Many important developments have since taken place and the originally planned book now contains additional recent contributions. So some of the frontiers and challenges we described at the beginning of 1987 are not quite the frontiers anymore. We have witnessed impressive technical advances, but many of the fundamental issues raised in our paper have not yet been resolved. So we felt it was better to leave the paper and its bibliography in their original version.

September 20, 1989

2.1. Introduction.

The subject of modeling and control of flexible structures has attracted wide interest over the past several years. The major driving force behind this interest is the very real possibility that large flexible space structures with stringent mission requirements will be launched within a decade. At present it is impossible to say with any certainty that current techniques are adequate to control the structures being envisioned. Thus, many practical and theoretical issues must be resolved before such structures are designed and flown.

Let us first turn to some practical questions. A first question concerns the mission requirements associated with a given structure; for example, rapidly slewing the structure over relatively large angles would imply large control energy which may excite the higher "models" leading to potential stability problems. It must be remembered that it is the control system itself which is the major source of energy input into the system, and thus looking only at the structure in open loop may not be meaningful. A second question concerns the type and positioning of sensors and actuators; reaction wheels, jets, or gyros may themselves be as massive as the total structure, thus completely changing the structural characteristics. In addition, each of these sensors or actuators has its own dynamics which may have to be taken into account. A third major question concerns inherent damping; i.e., damping inherent in the overall structure, something that everyone seems to agree will be present but no one seems able to model satisfactorily. For further insight into some of the practical issues, the reader is referred to Johnson [67].

Putting aside for the moment the questions raised in the previous paragraph, we now concentrate on some of the control models that have been proposed for flexible structures. A rather long body of literature (cf. Balas [3]) exists that essentially considers the flexible structure to be modeled by a finite-dimensional linear system of relatively high order. This is justified by arguing that either modal approximations or finite element methods will ultimately be used in calculations, and that the resulting models are as accurate as any other available. Further justification for these models is that modal or finite element techniques have been used successfully to analyze such complex structures as bridges and dams. Hence, the large finite-dimensional models are building upon a proven approach for certain structural problems. In parallel to (some would say orthogonal to) the approach just described, there is another body of literature which treats questions of estimation, identification, and control of partial differential equation models with the goal of analyzing and synthesizing controllers. Although the major portion of this literature treats parabolic systems, recent interest has centered around the problem of stabilization of hyperbolic systems. The pertinence of these models to the problem of control of large flexible structures is motivated by the following argument. Ideally, it would be desirable if the control system for a large flexible structure could be designed so that the rigid body modes could be precisely controlled while at the same time assuring that the flexibility does not degrade performance, or even worse cause failure of the structure. This would indicate that if the rigid and flexible modes are completely decoupled and if the controller at least assures the stability of the flexible modes, then the problem of controlling flexible structures is solvable. Thus both modeling approaches are pertinent in considering control questions for large flexible structures, and each has its drawbacks and advantages. It can be argued that the finite-dimensional approach may, on the one hand, mask certain fundamental problems (such as the effect of small delays on stability) and, on the other hand, introduce certain open questions that may be artificial (such as the spillover resulting from an exact design based on an a priori approximation of the system).

In this paper we will stress the viewpoint that partial differential equation (PDE) models must be used to provide insight into difficulties that may be encountered in controlling high performance large flexible space structures. We will review the existing results pertinent to stabilization and sensor-actuator placement for hyperbolic PDE's and will point out a number of open mathematical problems motivated by engineering considerations related to real sensors and actuators. At the present state of our knowledge the stabilization results regarding PDE models are limited to simple structures, for example, strings, beams, and plates. As such, this body or work is open to the criticism that "if you can't do much with simple structures why waste time with the PDE approach since real space structures will be very complex?" The answer is that if major difficulties arise in stabilizing and controlling simple structures, it is unrealistic to believe that these difficulties will disappear for complex structures. Furthermore, constructive techniques based on assembling simple free-free substructures into very complex structures already exist in the structural engineering literature (cf. Piché [108]). Thus any basic work on simple structures could be incorporated in an analogous way into the study of complex large flexible space structures.

In this paper we shall not discuss the linear quadratic problem and the construction of dynamical observers since extensive theoretical results already exist (cf. Lukes and Russell [97], Lions [89], Lasiecka and Triggiani [78]–[85]). The remaining important issues are

(i) to verify the associated stabilizability or the detectability hypotheses and

(ii) to make sense of the associated Riccati equations and study the uniqueness of solutions.

In addition we will not discuss identification problems or numerical methods. Here we will mainly be concerned with the stabilization problem.

2.2. Modeling.

In this section we give modeling equations for simple flexible structures from both finite element and PDE viewpoints. It is shown that the two models are "formally" equivalent. We then discuss inherent damping and certain classes of sensor and actuator models.

2.2.1. Modeling Equations. Our starting point is the general finite-dimensional rigid body model which is of the form

$$\bar{M}_{rr}\ddot{\bar{q}}_r = \bar{u}_r \tag{2.1}$$

where $\bar{M}_{rr}$ is a symmetric positive definite matrix, $\bar{q}_r$ is the vector of *rigid coordinates* and $\bar{u}_r$ is the vector of forces and torques applied to the structure.

In the structural engineering literature (cf., for instance, Hughes [65, §2], and Johnson [67]), the flexibility in the model is usually obtained by adding a vector $\bar{q}_e$ of elastic coordinates arising from a finite element approximation. A coupled set of equations is obtained

$$\bar{M}_{rr}\ddot{\bar{q}}_r + \bar{M}_{re}\ddot{\bar{q}}_e = \bar{u}_r \tag{2.2}$$

$$\bar{M}_{er}\ddot{\bar{q}}_r + \bar{M}_{ee}\ddot{\bar{q}}_e + \bar{C}_{ee}\dot{\bar{q}}_e + \bar{K}_{ee}\bar{q}_e = \bar{u}_e \tag{2.3}$$

where the new matrices verify the properties

(2.4) $$\bar{M}_{er}^T = \bar{M}_{re}$$

(2.5) $$\bar{M}_{ee}^T = \bar{M}_{ee} > 0 \quad \text{(positive definite)}$$

(2.6) $$\bar{C}_{ee}^T = \bar{C}_{ee} \nless 0 \quad \text{(not negative definite)}$$

(2.7) $$\bar{K}_{ee}^T = \bar{K}_{ee} > 0 \quad \text{(positive definite).}$$

For a complete description of the above model and the underlying assumptions the reader is referred to standard structural engineering texts (e.g., Clough and Penzien [22] or Meirovitch [104]). Introducing an appropriate control matrix F, the final model is of the form

(2.8) $$\bar{M}\ddot{\bar{q}} + \bar{C}\dot{\bar{q}} + \bar{K}\bar{q} = \bar{F}u$$

where

(2.9) $$\bar{q} = \begin{bmatrix} \bar{q}_r \\ \bar{q}_e \end{bmatrix}, \qquad u = \begin{bmatrix} \bar{u}_r \\ \bar{u}_e \end{bmatrix}$$

(2.10) $$\bar{M} = \begin{bmatrix} \bar{M}_{rr} & \bar{M}_{re} \\ \bar{M}_{er} & \bar{M}_{ee} \end{bmatrix}, \qquad \bar{K} = \begin{bmatrix} 0 & 0 \\ 0 & \bar{K}_{ee} \end{bmatrix}$$

(2.11) $$\bar{C} = \begin{bmatrix} 0 & 0 \\ 0 & \bar{C}_{ee} \end{bmatrix}, \qquad \bar{F} = \begin{bmatrix} \bar{F}_{rr} & \bar{F}_{re} \\ \bar{F}_{er} & \bar{F}_{ee} \end{bmatrix}$$

and usually the matrices $\bar{M}$ and $\bar{K}_{ee}$ are symmetric and positive definite.

The corresponding partial differential equation model is "formally" equivalent to equation (2.8) except for the vector of elastic coordinates $\bar{q}_e$ which becomes a vector of functions $\tilde{q}_e$ defined on the structure (space domain) Ω. In particular, the matrices $\bar{M}$, $\bar{C}$, $\bar{K}$, and $\bar{F}$ are replaced by operators $\tilde{M}$, $\tilde{C}$, $\tilde{K}$, and $\tilde{F}$ defined on appropriate spaces of functions on Ω:

(2.12) $$\tilde{M}\ddot{\tilde{q}} + \tilde{C}\dot{\tilde{q}} + \tilde{K}\tilde{q} = \tilde{F}\tilde{u}$$

where

(2.13) $$\tilde{q} = \begin{bmatrix} \tilde{q}_r \\ \tilde{q}_e \end{bmatrix},$$

(2.14) $$\tilde{M} = \begin{bmatrix} \tilde{M}_{rr} & \tilde{M}_{re} \\ \tilde{M}_{er} & \tilde{M}_{ee} \end{bmatrix}, \quad \tilde{K} = \begin{bmatrix} 0 & 0 \\ 0 & \tilde{K}_{ee} \end{bmatrix}, \quad \tilde{C} = \begin{bmatrix} 0 & 0 \\ 0 & \tilde{C}_{ee} \end{bmatrix}$$

(2.15) $$\tilde{F} = \begin{bmatrix} \tilde{F}_{rr} & \tilde{F}_{re} \\ \tilde{F}_{er} & \tilde{F}_{ee} \end{bmatrix}, \qquad \tilde{u} = \begin{bmatrix} \tilde{u}_r \\ \tilde{u}_e \end{bmatrix}.$$

A more precise description of F will be given in §2.2.3.

In order to illustrate the above abstract formulation, we consider the following examples.

Example 2.2.1. (*Free-free beam.*) Consider a thin elastic beam of length $2L > 0$ in free space away from gravity fields. Assume that its mass per unit length $\gamma > 0$ and its moment of inertia $I > 0$ are constant. Assume that the initial conditions and applied forces are such that the motion takes place in a plane. Denote by $y(x,t)$ the in-plane lateral displacement of the beam at position x and time t. (See Fig. 2.1.) The Euler–Bernoulli beam equation is

$$\gamma \frac{\partial^2 y}{\partial t^2} + EI \frac{\partial^4 y}{\partial x^4} = 0, \quad t > 0, \quad -L < x < L. \tag{2.16}$$

The initial conditions are

$$\begin{cases} y(x,0) = y^0(x), & -L < x < L \\ \dfrac{\partial y}{\partial x}(x,0) = y^1(x), & -L < x < L \end{cases} \tag{2.17}$$

for some appropriate functions y^0 and y^1 and the free-free boundary conditions are

$$\begin{cases} EI \dfrac{\partial^2 y}{\partial x^2}(-L,t) = EI \dfrac{\partial^2 y}{\partial x^2}(L,t) = 0, & t > 0, \\ EI \dfrac{\partial^3 y}{\partial x^3}(-L,t) = EI \dfrac{\partial^3 y}{\partial x^3}(L,t) = u_0(t), & t > 0 \end{cases} \tag{2.18}$$

where $u_0(t)$ is the control.

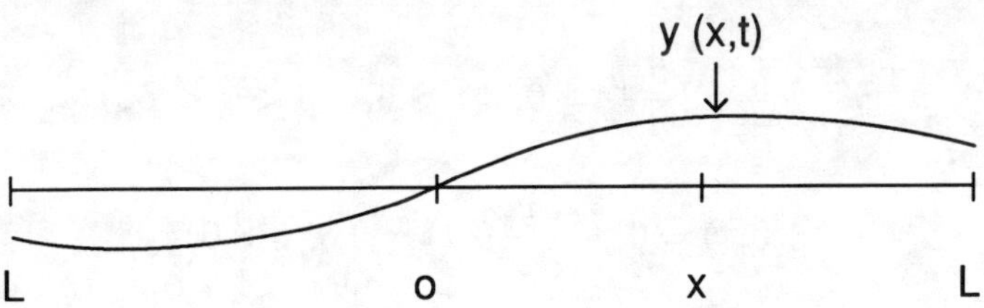

FIG. 2.1. *Lateral displacement of a thin beam.*

Equations (2.16) and (2.18) can be combined to obtain a more compact representation. Denote by W the Sobolev space $H^2(-L,L)$. Multiply both sides of (2.16) by an arbitrary function v in W and integrate over the interval $(-L,L)$:

$$\int_{-L}^{L} \gamma \frac{\partial^2 y}{\partial t^2}(x,t) v(x) dx + \int_{-L}^{L} EI \frac{\partial^4 y}{\partial x^4}(x,t) v(x) dx = 0. \tag{2.19}$$

Integrating by parts and using the boundary conditions, the second term becomes

(2.20) $$\int_{-L}^{L} EI\frac{\partial^2 y}{\partial x^2}(x,t)\frac{d^2 v}{dx^2}(x)dx + u_0(t)[v(L) - v(-L)].$$

Let δ_L and δ_{-L} be respectively the Dirac measure at L and $-L$

(2.21) $$\langle \delta_L, v\rangle_W = v(L), \quad \langle \delta_{-L}, v\rangle_W = v(-L)$$

where $\langle \cdot,\cdot\rangle_W$ is the duality pairing between W' and W. Let A be the continuous linear operator

(2.22) $$\langle Au, v\rangle_W = \int_{-L}^{L} EI\frac{d^2 u}{dx^2}(x)\frac{d^2 v}{dx^2}(x)dx.$$

and let $\dot{y}(t)$ and $\ddot{y}(t)$ be the functions $x \to \partial y(x,t)/\partial t$ and $x \to \partial^2 y(x,t)/\partial t^2$.

With the above notation, system (2.16)–(2.18) becomes

(2.23) $$\begin{cases} \gamma\ddot{y}(t) + Ay(t) + (\delta_L - \delta_{-L})u_0(t) = 0 \\ y(0) = y^0, \qquad \dot{y}(0) = y^1. \end{cases}$$

The first line of the above equation is to be interpreted with values in the topological dual space W' of W.

If we compare the last equation with (2.12)–(2.15), the operator A corresponds to K_{ee} δ_L, δ_{-L} to F_{ee} and γI (I the identity operator) to M_{ee}. The construction given in Example 2.1.1 is quite general and applies to many different situations.

Example 2.2.2. (*Clamped beam.*) We consider a clamped beam of length L with the same characteristics as in Example 2.1.1. The equations describing its deflection are

(2.24) $$\gamma\frac{\partial^2 y}{\partial t^2} + EI\frac{\partial^4 y}{\partial x^4} = 0, \quad t > 0, \quad 0 < x < L,$$

(2.25) $$y(0,t) = 0, \quad \frac{\partial y}{\partial x}(0,t) = 0, \quad t > 0,$$

(2.26) $$EI\frac{\partial^3 y}{\partial x^3}(L,t) = u_0(t) \text{ (force)}, \quad -EI\frac{\partial^2 y}{\partial x^2}(L,t) = u_1(t) \text{ (torque)}$$

and the initial conditions (2.17) for $0 < x < L$. This system can be rewritten in the following way

(2.27) $$\begin{cases} \gamma\ddot{y}(t) + Ay(t) + \delta_L u_0(t) + \delta_L' u_1 = 0 \quad \text{in } W' \\ y(0) = y^0, \qquad \dot{y}(0) = y^1 \end{cases}$$

where $\delta_{L'}$ is the distributional derivative of the Dirac measure.

(2.28) $$\langle \delta_L', v\rangle_W = \frac{dv}{dx}(L)$$

and the space W is now chosen as the subspace

$$W = \left\{ v \in H^2(0,L) : v(0) = 0, \frac{dv}{dx}(0) = 0 \right\}. \tag{2.29}$$

The last classical and simple example is the bar in torsion.

Example 2.2.3. (*Torsion of the beam.*) Consider a cylindrical beam of length $L > 0$. Denote by $y(x,t)$ the rotational displacement at position x and time t. Assume that its mass per unit length $\gamma > 0$, its torsional stiffness GJ, and radius of gyration ρ are constant.

The standard equations are

$$\gamma\rho^2 \frac{\partial^2 y}{\partial t^2} - GJ \frac{\partial^2 y}{\partial x^2} = 0, \quad t > 0, \quad 0 < x < L; \tag{2.30}$$

the initial conditions are

$$y(x,0) = y^0(x), \quad \frac{\partial y}{\partial x}(x,0) = y^1(x), \quad 0 < x < L, \tag{2.31}$$

for some appropriate functions y^0 and y^1; the boundary conditions are

$$GJ\frac{\partial y}{\partial x}(0,t) = u_0(t), \quad -GJ\frac{\partial y}{\partial x}(L,t) = u_1(t), \quad t > 0. \tag{2.32}$$

Again the system of equations (2.30)–(2.32) can be put in the more compact form

$$\begin{cases} \gamma\rho^2\ddot{y} + Ay + \delta_0 u_0(t) + \delta_L u_1(t) = 0, & t > 0, \\ y(0) = y^0, \qquad \dot{y}(0) = y^1 \end{cases} \tag{2.33}$$

where the first line of the above equation is to be interpreted in the topological dual of the Sobolev space $H^1(0,L)$ (functions on $(0,L)$ with a square integrable distributional derivative) and the operator

$$\langle Av, w\rangle_W = GJ \int_0^L \frac{dv}{dx}(x)\frac{dw}{dx}(x)dx. \tag{2.34}$$

Considerations analogous to the ones in Example 2.2.2 can be used to obtain an equation of the form (2.12) for the clamped beam in torsion.

The second-order abstract model (2.12) can also be represented as a first-order model by introducing the variables

$$y_1(t) = \tilde{q}(t), \qquad y_2(t) = \dot{\tilde{q}}(t) \tag{2.35}$$

and rewriting equation (2.12) in the form

$$\begin{bmatrix} I & 0 \\ 0 & \tilde{M} \end{bmatrix} \begin{bmatrix} \dot{y}_1 \\ \dot{y}_2 \end{bmatrix} = \begin{bmatrix} I & 0 \\ -\tilde{K} & -\tilde{C} \end{bmatrix} \begin{bmatrix} y_1 \\ y_2 \end{bmatrix} + \begin{bmatrix} 0 \\ \tilde{F} \end{bmatrix} \tilde{u}. \tag{2.36}$$

In the previous examples M is a bounded invertible symmetric positive operator. For simplicity, we assume that from now on we can rewrite (2.36) and choose M to be the identity. Thus, we shall consider systems of the form

$$\ddot{y} + G\dot{y} + Ay = Fu \tag{2.37}$$

or

$$\begin{bmatrix} \dot{y}_1 \\ \dot{y}_2 \end{bmatrix} = \begin{bmatrix} 0 & I \\ -A & -G \end{bmatrix} \begin{bmatrix} y_1 \\ y_2 \end{bmatrix} + \begin{bmatrix} 0 \\ F \end{bmatrix} u. \tag{2.38}$$

Let V and H be two Hilbert spaces. Identify the elements of the topological dual H' of H with those of H and assume that we have the classical chain of continuous injections

$$V \to H \equiv H' \to V'.$$

Let

$$A : V \to V' \tag{2.39}$$

be a continuous symmetric linear operator from a Hilbert space V into its dual space V'. Assume that it is V-H coercive

$$\exists \lambda \in \mathbb{R}, \quad \exists \alpha > 0 \quad \text{such that } \langle Av, v\rangle + \lambda \|v\|_H^2 \geq \alpha \|v\|_V^2. \tag{2.40}$$

Further assume that

$$F : U \to V' \tag{2.41}$$

is a continuous linear operator from an appropriate Hilbert space U of controls $u(t)$ into V' and that G is a damping operator that will be specified in the next section. In some cases the symmetry hypothesis on A can be relaxed.

2.2.2. Inherent Damping. By inherent damping we mean all energy-dissipation mechanisms that are inherent in the structure itself when placed in a vacuum or near vacuum. This includes material damping arising from physical properties of vibrating solids as well as sliding friction at joints and boundaries, but does not include damping that may be introduced by means of active controls.

We take as a starting point the model (2.38), i.e.,

$$\begin{bmatrix} \dot{y}_1 \\ \dot{y}_2 \end{bmatrix} = \begin{bmatrix} 0 & I \\ -A & -G \end{bmatrix} \begin{bmatrix} y_1 \\ y_2 \end{bmatrix} + \begin{bmatrix} 0 \\ F \end{bmatrix} u. \tag{2.42}$$

Let $u = 0$ in (2.42) and consider the following eigenvalue problem: find $\lambda \in \mathbb{C}$ for which there exists a pair of functions $(\varphi, \Psi) \neq (0,0)$ such that

(2.43)
$$\begin{bmatrix} 0 & I \\ -A & -G \end{bmatrix}\begin{bmatrix} \varphi \\ \Psi \end{bmatrix} = \lambda \begin{bmatrix} \varphi \\ \Psi \end{bmatrix}.$$

This problem is completely equivalent to that of finding $\varphi \neq 0$ such that

(2.44)
$$A\varphi + \lambda G\varphi + \lambda^2\varphi = 0.$$

This positive symmetric operator A is often used to obtain the λ's in terms of the eigenvalues and eigenvectors of A. The eigenvalues of A are all real, positive, or zero. They will be denoted

$$0 \leq \omega_1^2 \leq \omega_2^2 \cdots.$$

The corresponding set of eigenvectors is written

(2.45)
$$\Psi = \{\Psi_n : n = 1, 2, \cdots\}.$$

Assume that Ψ forms an orthonormal basis of H:

(2.46)
$$(\Psi_n, \Psi_m)_H = \begin{cases} 1, & m = n \\ 0, & m \neq n \end{cases}$$

(2.47)
$$\langle A\Psi_n, \Psi_m\rangle_V = \begin{cases} \omega_n^2, & m = n \\ 0, & m \neq n. \end{cases}$$

The set Ψ of eigenvectors is contained in $D(A) \subset V$ and its elements are orthogonal. Further assume that the set Ψ is complete in V. For instance, this occurs when A has a bounded inverse A^{-1} from H to H which is compact (cf. Taylor [132, Thm 6.41-A, p. 343]).

In view of the above assumptions, an eigenvector ϕ in equation (2.44) can be written

(2.48)
$$\phi(x) = \sum_{n=1}^{\infty} \phi_n \Psi_n(x), \qquad \phi_n \in \mathbb{R} \quad \text{(field of real numbers)}.$$

Using this representation, (2.44) reduces to

(2.49)
$$\omega_n^2 + \lambda \sum_{m=1}^{\infty} (G\phi_m, \phi_n)_H + \lambda^2 = 0 \quad \forall n \geq 1.$$

When G is a multiple of the identity operator I,

(2.50)
$$G = 2\gamma I, \qquad \gamma \geq 0$$

the system of equations (2.49) is completely decoupled:

(2.51)
$$\omega_n^2 + 2\lambda\gamma + \lambda^2 = 0 \quad \forall n \geq 1.$$

Thus

(2.52) $$\lambda_n = -\gamma \pm \sqrt{\gamma^2 - \omega_n^2} \quad \forall n \geq 1.$$

When $\gamma = 0$ (the undamped case)

(2.53) $$\lambda_n = \pm i\omega_n$$

where $i = \sqrt{-1}$. When $\gamma^2 \leq \omega_1^2$ (viscous damping)

(2.54) $$\lambda_n = -\gamma \pm i\sqrt{\omega_n^2 - \gamma^2} \quad \forall n \geq 1.$$

When the damping operator G is of the form (2.50) with $\gamma > 0$ the eigenvalues all have negative real parts equal to $-\gamma$. It can also be shown that the energy associated with the system

(2.55) $$E(y(t), \dot{y}(t)) = \tfrac{1}{2}\left[\|\dot{y}(t)\|_H^2 + \langle Ay(t), y(t)\rangle_V\right]$$

decays at a uniform exponential rate as $t \to \infty$. The same type of result can be obtained for damping operators which satisfy the following conditions:

(2.56) $$G : H \to H \quad \text{linear and continuous}$$

(2.57) $$\exists \alpha > 0, \quad \forall v \in H, \quad (Gv, v)_H \geq \alpha \|v\|_H^2.$$

This is the so-called *uniform exponential stability.*

Another interesting type of damping is the notion of *weak damping*, which is characterized by damping operators with the following properties

(2.58) $$G : H \to H \quad \text{linear and continuous}$$

(2.59) $$\forall v \in H, \quad (Gv, v)_H \geq 0.$$

This type of damping has been studied by Dafermos [31] and Russell [114]. *Weakly damped* systems correspond to the situation where the damping operator G is not invertible (that is, (2.57) does not hold). This terminology was introduced by Dafermos [31]. Most (but not all) applications where weak damping occurs involve a compact operator G and conclusions of the type

(2.60) $$\lim_{t\to\infty} E(y(t), \dot{y}(t)) = 0.$$

Russell [114] studied the decay rates of weakly damped systems. He showed that *uniform decay rates are not to be expected if G is compact while the unit ball in H is not.* However, he was able to obtain decay rates for the energy (2.55) provided that the compact operator G and the initial state $(y(0), \dot{y}(0))$ satisfy "certain strong (but not unrealistic) requirements." He also obtained some results in the case where G is not compact but also not invertible. He claims that this occurs in connection with certain frictionally coupled elastic systems (cf. Russell [114] for additional details).

Models (2.56)–(2.59) for the damping operator G do not satisfactorily describe the inherent damping of flexible structures. Although not universally accepted, empirical studies (cf. Skudrzyk [128], Kolousek [68], Clough and Penzien [22], Hughes [65]) seem to indicate that the natural modes of vibration have damping rates which are roughly proportional to the frequency of vibration. This damping behavior is sometimes called "structural damping." A number of ad hoc models exhibiting behavior of this type have been proposed in the engineering literature (see Piché [108] for a discussion of "Complex Stiffness" p. 68 and "Signum (Hysteretic) Damping," p. 70). These models, however, yield, respectively, spurious unstable eigenvalues or noncausal systems, thus they do not seem useful for control purposes. In light of the above considerations, Chen and Russell [19] have suggested a damping operator G of the form

(2.61) $$G = 2\gamma A^{1/2}, \qquad \gamma > 0,$$

where $A^{1/2}$ is the positive square root operator of A. It is a self-adjoint operator which is linear and continuous

(2.62) $$A^{1/2} : D(A^{1/2}) = V \to H.$$

As a result $A^{1/2}$ is also linear and continuous

(2.63) $$A^{1/2} : H \to D(A^{1/2})' = V'.$$

By using interpolation theory $A^{1/2}$ is also linear and continuous on intermediate spaces

(2.64) $$A^{1/2} : [V, H]_{1/2} \to [H', V']_{1/2} = [V, H]'_{1/2}.$$

The space $[V, H]_{1/2}$ is a Hilbert space which lies halfway between V and H with continuous dense injections

(2.65) $$V \subset [V, H]_{1/2} \subset H \equiv H' \subset [V, H]'_{1/2} \subset V'.$$

In particular,

(2.66) $$\langle A^{1/2}u, v\rangle_{[V,H]_{1/2}} = \langle A^{1/2}v, u\rangle_{[V,H]_{1/2}} \quad \forall u, v$$

(2.67) $$\langle A^{1/2}u, u\rangle_{[V,H]_{1/2}} \geq 0 \quad \forall u, \in [V, H]_{1/2}.$$

If, in addition, A is coercive, $A^{1/2}$ is also coercive.

The fact that A is a differential operator does not at all insure that $A^{1/2}$ is also a differential operator. In general $A^{1/2}$ is a nonlocal operator. This issue is discussed in Chen and Russell [19]. They point out that when A

is approximated by a symmetric positive definite matrix $\bar{A}$, the symmetric positive definite square root $\bar{A}^{1/2}$ of $\bar{A}$ may be readily computed and the corresponding finite-dimensional system

$$\ddot{\bar{y}} + 2\gamma\bar{A}^{1/2}\dot{\bar{y}} + \bar{A}\bar{y} = 0 \tag{2.68}$$

is then available for computational purposes. The outcome of this construction is that the amplitudes of the normal modes of vibration are attenuated at rates that are proportional to the oscillation frequencies of the matrix $\bar{A}$ in good agreement with experimental observation. Structural engineers obtain the same result (both qualitatively and quantitatively) by introducing the damping directly on the transfer function model. For modal approximation of A, this is probably a good way to go. However, for a finite element approximation of A, this may raise some convergence questions connected with the approximation of $A^{1/2}$.

In order to illustrate the above abstract considerations, we set $G = 2\gamma A^{1/2}$ in the eigenvalue problem (2.43); find $\phi \neq 0$ such that

$$A\phi + \lambda 2\gamma A^{1/2}\phi + \lambda^2\phi = 0. \tag{2.69}$$

Denote by $0 \leq \omega_1 \leq \omega_2 \leq \cdots$ the ordered sequence of eigenvalues of the operator $A^{1/2}$ where an eigenvalue with multiplicity m greater than one is repeated m times. Let

$$\Xi = \{\xi_n | n = 1, 2, \cdots\} \tag{2.70}$$

be the corresponding set of eigenvectors. Assume that Ξ forms an orthonormal basis of H:

$$(\xi_n, \xi_m)_H = \begin{cases} 1, & m = n \\ 0, & m \neq n \end{cases} \tag{2.71}$$

$$\langle A^{1/2}\xi_n, \xi_m\rangle_{[V,H]_{1/2}} = \begin{cases} \omega_n & \text{if } m = n \\ 0 & \text{if } m \neq n. \end{cases} \tag{2.72}$$

Further assume that the set Ξ is complete in $[V, H]_{1/2}$.

Formally

$$A\xi_n = A^{1/2}A^{1/2}\xi_n = A^{1/2}\omega_n\xi_n = \omega_n^2\xi_n \tag{2.73}$$

and the set $\{\omega_n^2 | n = 1, 2, \cdots\}$ belongs to the set of eigenvalues of the operator A.

In particular,

$$(2.74)\quad \begin{aligned} \langle A\xi_n, \xi_m\rangle_V &= \left(A^{1/2}\xi_n, A^{1/2}\xi_m\right)_H \\ &= \omega_m\langle A^{1/2}\xi_n, \xi_m\rangle_{[V,H]_{1/2}} \\ \langle A\xi_n, \xi_m\rangle_V &= \left\{ \begin{array}{ll} \omega_n^2, & m = n \\ 0, & m \neq n \end{array} \right\}. \end{aligned}$$

In view of the above properties, we can go back to equation (2.68) and write the eigenvector ϕ in terms of the eigenvectors Ξ of $A^{1/2}$

$$(2.75)\quad \phi(x) = \sum_{n=1}^{\infty} \phi_n \xi_n(x), \qquad \phi_n \in \mathbb{R}.$$

Equation (2.68) is equivalent to the following set of equations: for all $n \geq 1$

$$(2.76)\quad \omega_n^2 + 2\gamma\lambda\omega_n + \lambda^2 = 0.$$

So for each $n \geq 1$

$$(2.77)\quad \lambda_n = \omega_n \left[-\gamma \pm \sqrt{\gamma^2 - 1}\right].$$

For lightly damped structures $(0 < \gamma < 1)$

$$(2.78)\quad \lambda_n = \omega_n \left[-\gamma \pm i\sqrt{1 - \gamma^2}\right].$$

Thus we see that under these conditions, the modes of vibration having damping rates which are proportional to the frequencies of vibration and this type of inherent damping is often used in finite-dimensional models.

Chen and Russell [19] have also looked at damping operators of a form less restrictive than $G = 2\gamma A^{1/2}$ or $G = 2\gamma A$. The reader is referred to their paper for detailed conditions on the operator G.

If the structure is made of a material modeled by the Voigt–Kelvin model (cf. Popov [109] and Clough and Penzien [22]) for viscoelasticity, the damping operator is

$$(2.79)\quad G = 2\gamma A, \qquad \gamma \geq 0.$$

Here equation (2.44) reduces to

$$(2.80)\quad A\phi + \lambda 2\gamma A\phi + \lambda^2\phi = 0.$$

By using the set of eigenvectors Ψ (2.45)–(2.47) associated with the operator A,

$$(2.81)\quad \phi(x) = \sum_{n=1}^{\infty} \phi_n \Psi_n(x)$$

and (2.80) is equivalent to the following set of equations: for all $n \geq 1$

$$\omega_n^2 + 2\gamma\lambda\omega_n^2 + \lambda^2 = 0. \tag{2.82}$$

So for each $n \geq 1$

$$\lambda_n^{\pm} = \omega_n \left[-\gamma\omega_n \pm \sqrt{(\gamma\omega_n)^2 - 1} \right]. \tag{2.83}$$

When ω_n goes to $+\infty$

$$\lambda_n^- \to -\infty \quad \text{and} \quad \lambda_n^+ \to -\frac{1}{2\gamma}. \tag{2.84}$$

(See Fig. 2.2.) This type of damping operator is covered by Chen and Russell [19].

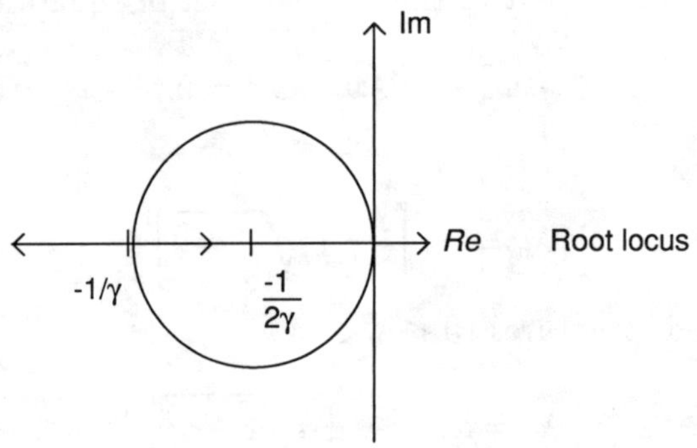

FIG. 2.2. *The root locus for the Voigt–Kelvin model.*

Other damping models are of the form

$$(\mu I + vA)\ddot{y}(t) + G\dot{y}(t) + Ay(t) + G_0(Ay)_t = 0 \tag{2.85}$$

where $\mu > 0$, $v \geq 0$, $G : V \to V'$,

$$G_0 : C_0(-\infty, 0; V') \to V', \qquad G_0\varphi = \int_{-\infty}^{0} g(\theta)\varphi(\theta)\, d\theta \tag{2.86}$$

where $C_0(-\infty, 0; V')$ is the space of continuous functions that go to zero at $-\infty$ and for Ay defined on $[0, T]$: for all t in $[0, T]$

$$(Ay)_t(\theta) = \left\{ \begin{array}{ll} (Ay)(t+\theta), & t+\theta \geq 0 \\ 0, & t+\theta < 0 \end{array} \right\}, \qquad -\infty < \theta \leq 0. \tag{2.87}$$

Such models are reviewed in Piché [108]. For instance,

$$v > 0, \quad G = 2\gamma A, \quad \gamma > 0 \quad \text{and} \quad G_0 = 0 \tag{2.88}$$

is what he calls the Higuchi, Bland, Lee damping; other examples involving the operator G_0 are the Fractional Derivative damping and the Biot damping, where $v = 0$ and $G = 0$. Models involving G_0 originate with Boltzmann's study of viscoelasticity. Recall that a fractional power of the operator A can be nonlocal.

If the operator G_0 is chosen as

$$G_0(Ay)_t = 2\gamma[Ay(t) - Ay(t-\tau)]/\tau \tag{2.89}$$

for some small $\tau > 0$ and $\gamma > 0$, we have an approximation to the Voigt damping, $2\gamma Ay(t)$ and it is not too surprising to obtain a stabilizing effect.

Thus, in addition to the memoryless damping mechanisms of (2.88) it is also possible to include certain damping mechanisms with memory into the general model (2.37). The form of the damping operator to be used in modeling large flexible dynamic structures is a current research topic since none of the operators discussed in this section seem to be completely adequate in all situations. The importance of the inherent damping lies in the fact that, as will be discussed in §2.3, the possibility of stabilizing the system using certain types of feedback configurations is critically dependent on the inherent damping.

2.2.3. Sensors and Actuators. This topic is too broad to present a complete analysis of all the engineering and mathematical questions associated with sensors and actuators. Thus, we will concentrate on some interesting cases which lend themselves to relatively simple mathematical analysis.

In §2.2.1 we have seen two examples of actuators: point forces and torques. In both cases, the general model was

$$\begin{cases} \ddot{y} + G\dot{y} + Ay = Fu \quad \text{in } V' \\ y(0) = y^0 \in V, \qquad \dot{y}(0) = y^1 \in H \end{cases} \tag{2.90}$$

where the control operator F is a continuous linear operator from a Hilbert space of controls U into the topological dual of a Hilbert space V

$$F : U \to V' \tag{2.91}$$

and A and G are appropriate continuous linear operators

$$A : V \to V', \qquad G : V \to V' \tag{2.92}$$

described in §§2.2.1 and 2.2.2.

Such sensors can roughly be described as devices which are "dual" to actuators. Let Y be a Hilbert space where the observations of system (2.90) will take place. Y will be referred to as the *observation space.* For the moment, consider a continuous linear operator

$$C : V \to Y$$

and associate with it the *observation*

$$z(t) = Cy(t).$$

The operator C will be referred to as the *observation* operator.

In practice, the space of controls U and the space of observations Y are finite-dimensional. When the observation or output is fed back to the control through the linear operator

$$K : Y \to U,$$

the resulting closed-loop operator becomes

$$FKC : V \to V'. \tag{2.93}$$

When both U and Y are finite-dimensional, K is a matrix, and F and C are compact operators. Then the composition $FKC : V \to V'$ is a compact operator.

2.2.3.1. Distributed or bounded observation and control operators. From an engineering viewpoint point, sensors and actuators (S/A) are useful idealizations that can never be achieved in practice. All real sensors will yield measurements of the variables being monitored, which are weighted averages over a portion of the spatial domain. In the same way, the control actions resulting from real actuators affect a distributed part of the spatial domain. This is often known as *distributed observation and control.* However, as we will discuss below, care must be taken in defining a mathematical notion of distributed operator. In particular, it would be more realistic to consider "concentrated" S/A rather than point S/A. By concentrated S/A we mean S/A whose observations and actions are really distributed but concentrated around a specific spatial location, i.e., weighted averages over small regions of the spatial domain.

Consider the following example with $H = L^2(\Omega)$ and control and observation operators that are continuous with H in place of V

$$F : U \to H \text{ (respectively, } C : H \to Y).$$

When U (respectively, Y) is the N-dimensional Euclidean space, the distributed control (respectively, observation) operator has a square integrable integral representation

$$(Fu, \phi)_H = \int_\Omega [f(x) \cdot u]\phi(x)dx$$

$$\left(\text{respectively, } C\phi = \int_\Omega c(x)\phi(x)dx\right)$$

where f (respectively, c) is an N-vector of square integrable functions over Ω, and $v \cdot w$ denotes the N-dimensional scalar product between two arbitrary N-vectors v and w. The composition FKC has the following integral representation

$$(FKC\varphi)(x) = \int_\Omega [f(x) \cdot Kc(y)]\varphi(y)dy$$

where

$$\int_\Omega \int_\Omega |f(x) \cdot Kc(y)|^2 dx\, dy < +\infty.$$

As a result FKC is a Hilbert–Schmidt operator and a fortiori FKC is compact.

Consider now Example 2.2.2 in §2.2.1 with $u_0(t) = 0$ and control through a torque actuator u_1 in (2.26). The corresponding control operator F is the derivative of the Dirac delta function at L, that is,

$$\langle \phi, Fu_1(t)\rangle = \frac{d\phi}{dx}(L)u_1(t) \quad \text{for all } \phi \text{ in } W.$$

The corresponding "distributed control operator" F_ρ would be of the form

$$\langle \phi, F_\rho u_1(t)\rangle = \int_0^L \rho(x)\frac{d\phi}{dx}(x)dx\, u_1(t) \quad \text{for all } \phi \text{ in } W$$

for some L^2-function ρ. F_ρ is not a continuous functional on $L^2(0, L)$. This illustrates the necessity of giving a more precise mathematical sense to the notion of "distributed" observation and control.

The integral representation of the control or observation operators cannot, however, be retained since any functional L on a Sobolev space $H^m(\Omega)$ has an integral representation of the form

$$L\phi = \sum_{|\alpha|=m} \int_\Omega \xi_\alpha(x) D^\alpha \varphi(x)dx$$

for some family $\{\xi_\alpha\}$ of L^2-function on Ω. Thus any point S/A can be represented in this form for some power m and the existence of an integral representation cannot be used as a criterion for the notion of a "distributed operator."

What seems important is how large this power m is when compared to the one of the Sobolev space in which the time-derivative of the state $y(t)$ lies. It is really this gap which is important mathematically and not the fact that the operator FKC is physically distributed. For the general model (2.90), it is natural to say that the control (respectively, observation) operator is *bounded* when the operator F (respectively, C) is H-continuous, that is, the linear map

$$F : U \to H \qquad (\text{respectively, } C : H \to Y). \tag{2.94}$$

is continuous. Otherwise, we shall say that F (respectively, C) is *unbounded.* It is to be emphasized that, in general, the space H is not necessarily $L^2(\Omega)$, but the space associated with the underlying semigroup on $V \times H$. In many examples H will be $L^2(\Omega)$, but this is not always the case.

In the sequel we shall use the notion of boundedness or more precisely H-continuity which is mathematically more suitable for our purpose.

2.2.3.2. Boundary and point control operators. Boundary control and observation operators do not belong to the class of "bounded operators" with finite-dimensional control and observation spaces (except in space dimension one where the boundary is composed of two points). Large classes of boundary controls are covered by the model (2.90)–(2.91), which is essentially based on variational techniques. We say that the control operator F is a *boundary control operator* if U is a space of traces on the boundary Γ or part of the boundary Γ. Similarly, we shall say that F is a *point control operator* if U is the N-dimensional Euclidean space and F is made up of N Dirac distributions or partial derivatives of Dirac distributions at N points in the domain Ω: in space dimension one

$$F(u_1(t), \cdots, u_N(t)) = \sum_{i=1}^{N} F_i \delta_{x_i}^{(n_i)} u_i(t)$$

where F_i is a matrix and $\delta_{x_i}^{(n_i)}$ is the n_ith spatial derivative of the Dirac distribution at the point $x_i \in \Omega$, $1 \leq i \leq N$.

More demanding types of boundary or point control will necessitate the use of the method of transposition and be modeled by an equation of the type

$$\text{(2.95)} \qquad \begin{cases} \ddot{y} + G\dot{y} + Ay = Fu(t) & \text{in } D(A^*)' \\ y(0) = y^0 \in H, & \dot{y}(0) = y^1 \in D(A^*)' \end{cases}$$

where now the operators

$$\text{(2.96)} \qquad F : U \to D(A^*)', \quad A : H \to D(A^*)', \quad G : H \to D(A^*)'$$

are linear and continuous but with values in the topological dual $D(A^*)'$ of the domain $D(A^*)$ of the adjoint operator of A

$$\text{(2.97)} \qquad D(A^*) = \{v \in V | A^*v \in H\}.$$

The domain $D(A^*)$ of A^* is contained in V, and we have the chain of spaces

$$\text{(2.98)} \qquad D(A^*) \subset V \subset H \equiv H' \subset V' \subset D(A^*)'.$$

The gain is in the fact that $D(A^*)'$ is a bigger space than V'. This will permit a larger family of boundary controls as we shall see in the examples.

Model (2.95), however, is not that different from model (2.90). When A is coercive

$$\text{(2.99)} \qquad \exists \alpha > 0 \text{ such that } \forall v \in V, \quad \langle Av, v \rangle_V \geq \alpha \|v\|_V^2$$

and the inverse $A^{-1/2}$ of the positive square root operator exists. By changing the variable y to

$$\bar{y}(t) = A^{1/2}y(t)$$

we obtain

$$(2.100)\qquad \begin{cases} \ddot{\bar{y}}(t) + A^{-1/2}GA^{1/2}\dot{\bar{y}}(t) + Ay(t) = A^{-1/2}Fu(t) & \text{in } V' \\ \bar{y}(0) = A^{-1/2}y^0 \in V, \qquad \dot{\bar{y}}(0) = A^{-1/2}y^1 \in H. \end{cases}$$

By noticing that

$$(2.101)\qquad \begin{cases} A : V \to V', \qquad A^{-1/2}F : U \to V' \\ A^{-1/2}GA^{1/2} : V \to V' \end{cases}$$

we again have model (2.90) with a new F and a new G.

Example 2.2.4. Let Ω be a bounded open domain in $\mathbb{R}^k$ and let Γ be its boundary. (See Fig. 2.3.) Consider the wave equation

$$(2.102)\qquad \frac{\partial^2 y}{\partial t^2}(x,t) - \Delta y(x,t) = 0, \qquad x \in \Omega, \quad t > 0$$

with boundary condition

$$(2.103)\qquad \frac{\partial y}{\partial n}(x,t) = u(x,t), \qquad x \in \Gamma, \quad t > 0$$

and initial conditions

$$(2.104)\qquad y(x,0) = y^0(x), \qquad \frac{\partial y}{\partial t}(x,0) = y^1(x), \quad x \in \Omega,$$

where Δ is the Laplacian on $\mathbb{R}^k$

$$(2.105)\qquad \Delta y(x,t) = \sum_{i=1}^{k} \frac{\partial^2 y}{\partial x_i^2}(x,t), \qquad x = (x_1, \cdots, x_k) \in \Omega,$$

$n(x)$ is the unit external normal to Ω at $x \in \Gamma$,

$$(2.106)\qquad \vec{n}(x) = (n_1(x), \cdots, n_k(x)), \qquad x \in \Gamma$$

and $\partial y/\partial n$ is the normal derivative to Ω at $x \in \Gamma$

$$(2.107)\qquad \frac{\partial y}{\partial n}(x,t) = \sum_{i=1}^{k} \frac{\partial y}{\partial x_i}(x,t)n_i(x).$$

The above system is modeled by (2.90) by choosing

$$(2.108)\qquad V = H^1(\Omega), \qquad H = L^2(\Omega).$$

$$(2.109)\qquad \langle Ay, v\rangle_V = \sum_{i=1}^{k} \int_\Omega \frac{\partial y}{\partial x_i}(x)\frac{\partial v}{\partial x_i}(x)dx \quad \forall v \in V$$

$$\langle Fu, v\rangle_V = \int_\Gamma u(x)v(x)dx \quad \forall v \in V. \tag{2.110}$$

Here $H^1(\Omega)$ is the Sobolev space of functions y in $L^2(\Omega)$ with distributional derivatives

$$\frac{\partial y}{\partial x_i} \in L^2(\Omega), \qquad 1 \le i \le k.$$

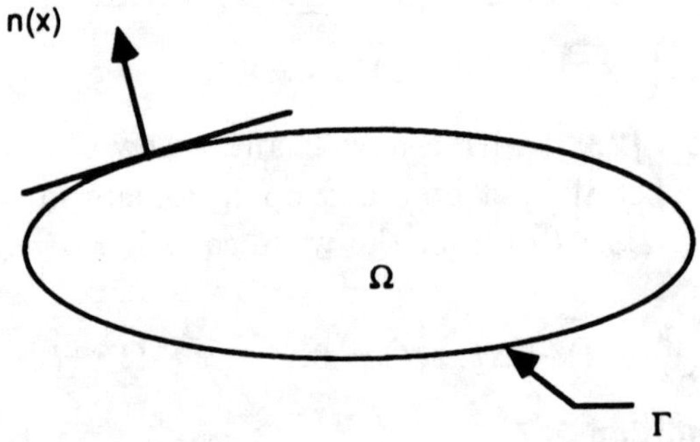

FIG. 2.3. *A bounded domain Ω with boundary Γ and unit normal $n(x)$.*

Example 2.2.5. With the notation of Example 2.2.4, consider the wave equation (2.104) with boundary condition

$$y(x,t) = u(x,t), \qquad x \in \Gamma, \quad t > 0 \tag{2.111}$$

and initial conditions (2.104). Here we choose

$$H = L^2(\Omega) \quad \text{and} \quad V = H_0^1(\Omega).$$

Let $D(A^*)$ be the domain of the operator A^* with respect to H:

$$D(A^*) = \{v \in H^2(\Omega) |\ v = 0 \text{ on } \Gamma\}. \tag{2.112}$$

By using integration by parts, system (2.102), (2.104), and (2.111) is equivalent to

$$\ddot{y} + Ay = Fu \quad \text{in } D(A^*)', \tag{2.113}$$

where

$$A : H \to D(A^*)' \quad \text{and} \quad F : U = L^2(\Gamma) \to D(A^*)' \tag{2.114}$$

are defined as

$$\langle Ay, v\rangle_{D(A^*)} = -(y, \Delta v)_H, \quad v \in D(A^*) \tag{2.115}$$

$$\langle Fu, v\rangle_{D(A^*)} = -\int_\Gamma u(x)\frac{\partial v}{\partial n}(x)d\Gamma, \quad v \in D(A^*). \tag{2.116}$$

So we are back to model (2.95).

Example 2.2.6. With the notation of Example 2.2.4, consider the wave equation (2.102) with the boundary conditions

$$\begin{aligned} y(x,t) &= u_0(x,t) \qquad \text{on } \Gamma_0, \quad t>0, \\ \frac{\partial y}{\partial n}(x,t) &= u_1(x,t) \quad \text{on } \Gamma_1, \quad t>0, \end{aligned} \tag{2.117}$$

where the boundary Γ of Ω is made up of two parts

$$\Gamma = \Gamma_0 \cup \Gamma_1. \tag{2.118}$$

(See Fig. 2.4.)

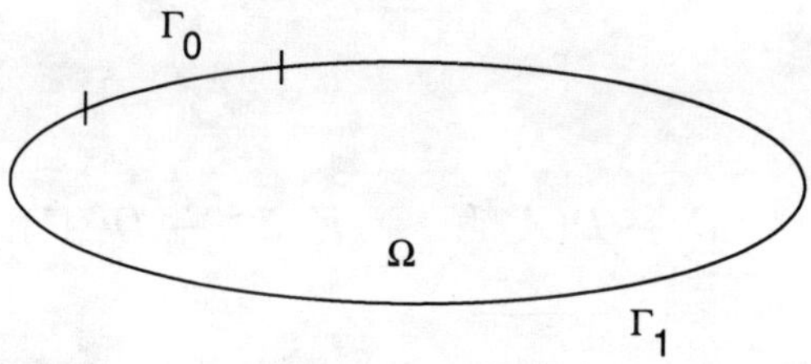

FIG. 2.4. *A bounded domain Ω with boundary Γ made up of two parts Γ_0 and Γ_1.*

Choose

$$\begin{cases} H = L^2(\Omega), \qquad V = \{v \in H^1(\Omega) | v|_{\Gamma_0} = 0\} \\ D(A^*) = \left\{ v \in H^2(\Omega) | v|_{\Gamma_0} = 0, \dfrac{\partial v}{\partial n}\Big|_{\Gamma_1} = 0 \right\} \end{cases} \tag{2.119}$$

By integration by parts, system (2.102), (2.104), and (2.117) becomes

$$\ddot{y}(t) + Ay(t) = F(u_0(t), u_1(t)) \tag{2.120}$$

where

$$A : H \to D(A^*)', \qquad F : L^2(\Gamma_0) \times L^2(\Gamma_1) \to D(A^*)' \tag{2.121}$$

are defined as

$$\begin{cases} \langle Ay, v\rangle_{D(A^*)} = -(y, \Delta v)_H, \qquad v \in D(A^*) \\ \langle F(u_0, u_1), v\rangle_{D(A^*)} = \displaystyle\int_{\Gamma_1} u_1(x)v(x)ds - \int_{\Gamma_0} u_0(x)\frac{\partial v}{\partial n}(x)ds, \end{cases} \tag{2.122}$$

v in $D(A^*)$. This problem is modeled by (2.95).

Example 2.2.7. Consider the elastic beam (2.24) with initial condition (2.25), but with the following new boundary conditions:

$$(2.123)\quad \begin{cases} -EI\,y(0,t) = u_0(t), & EI\dfrac{\partial y}{\partial x}(0,t) = u_1(t) \\ EI\dfrac{\partial^2 y}{\partial x^2}(L,t) = u_2(t), & -EI\dfrac{\partial^3 y}{\partial x^3}(L,t) = u_3(t). \end{cases}$$

Here

$$(2.124)\quad H = L^2(0,L), \qquad V = \left\{ v \in H^2(0,L) | v(0) = \frac{dv}{dx}(0) = 0 \right\}.$$

We introduce

$$(2.125)\quad D(A^*) = \left\{ v \in H^4(0,L) | v(0) = \frac{dv}{dx}(0) = \frac{d^2v}{dx^2}(L) = \frac{d^3v}{dx^3}(L) = 0 \right\}$$

and the operators

$$(2.126)\qquad A : H \to D(A^*)', \qquad F : \mathbb{R}^4 \to D(A^*)'$$

defined by

$$\langle Ay, v\rangle_{D(A^*)} = \int_\Omega y(x)\frac{d^4v}{dx^4}(x)dx$$

$$(2.127)\quad \langle F(u_0(t), u_1(t), u_2(t), u_3(t)), v\rangle_{D(A^*)}$$
$$= -u_0(t)\frac{d^3v}{dx^3}(0) + u_1(t)\frac{d^2v}{dx^2}(0) + u_2(t)\frac{dv}{dx}(L) - u_3(t)v(L)$$

Then system (2.24), (2.25), (2.123) can be put in the form (2.95).

The same exercise can be repeated for different sets of boundary conditions at the end of the beam (cf. Example 2.2.2 with nonhomogeneous boundary conditions).

2.2.3.3. Smoothness of solutions. Going to more general classes of control operators is quite natural. The price to pay is that the associated solution will be rougher and rougher. Fortunately, recent regularity results are now available (cf. Lions [90], Chapter 1 in this volume).

2.2.3.4. Observation operators. The class of observation operators

$$(2.128)\qquad C : V \to Y$$

introduced at the beginning of §2.2.3 is not necessarily limited to continuous maps on H or V. It can be weakened to accommodate point or boundary observations which require more smoothness on the state variable

$$(2.129)\qquad C : D(A^n) \to Y, \qquad n \geq 0$$

$$D(A^n) \subset D(A^{n-1}) \subset \cdots \subset D(A) \subset V \subset H. \tag{2.130}$$

By analogy with the preceding terminology for control operators we shall say that C is a *boundary observation operator* if Y is a space of traces on the boundary Γ or part of the boundary Γ (cf. Lions and Magenes [93], [94]). Similarly, C is a *point observation operator* if Y is the N-dimensional Euclidean space and C is made up of N evaluations of the state or partial derivatives of the state at N points in the domain Ω. In space dimension one

$$C\phi = \sum_{i=1}^{N} C_i D^{n_i} \phi(x_i) \tag{2.131}$$

where C_i is a matrix and $D^{n_i}\phi(x_i)$ is the n_ith spatial derivative of the function ϕ at the point $x_i \in \Omega$, $1 \leq i \leq N$. In space dimensions greater or equal to two a simple point observation in the domain is obtained by going from the space V to the domain $D(A)$ of the operator A with respect to H

$$D(A) = \{v \in V | Av \in H\} \tag{2.132}$$

$$C : D(A) \to Y \tag{2.133}$$

or even to the domains of A^2, A^3, etc. ..., in order to make sense of the maps

$$v \to v(x_0), \qquad v \to \frac{\partial v}{\partial x}(x_0), \text{ etc. } \ldots \tag{2.134}$$

Many interesting observation operators can be simply obtained by constructing the dual F^* to the control operators introduced in §§2.2.1 and 2.2.3.2. So we systematically give the observation operators which are the dual of the control operators found in Examples 2.2.1 through 2.2.7.

Example 2.2.8. Here the observation could be the velocities

$$z_1(t) = \frac{\partial y}{\partial t}(-L, t), \tag{2.135}$$

$$z_2(t) = \frac{\partial y}{\partial t}(L, t). \tag{2.136}$$

This corresponds to the observation operator

$$C\phi = \phi(L) - \phi(-L) \tag{2.137}$$

applied to the velocity $\dot{y}(t)$

$$z(t) = (z_2(t), z_1(t)) = C\dot{y}(t). \tag{2.138}$$

Example 2.2.9. Choose the observation

(2.139) $$z(t) = \left(\frac{\partial y}{\partial t}(L,t), \frac{\partial^2 y}{\partial x\, \partial t}(L,t)\right)$$

which is equivalent to

(2.140) $$z(t) = C\dot{y}(t)$$

where

(2.141) $$C\phi = \left(\phi(L), \frac{\partial \phi}{\partial x}(L)\right).$$

Example 2.2.10. Choose the observation

(2.142) $$z(t) = \left(\frac{\partial y}{\partial t}(0,t), \frac{\partial y}{\partial t}(L,t)\right)$$

which is equivalent to (2.140) with

(2.143) $$C\phi = (\phi(0), \phi(L)).$$

Example 2.2.11. Choose the observation

(2.144) $$z(t) = \left.\frac{\partial y}{\partial t}\right|_{\Gamma}$$

which is equivalent to (2.140) where

(2.145) $$\phi \mapsto C\phi = \varphi|_{\Gamma} : V \to H^{1/2}(\Gamma)$$

is the trace operator.

Example 2.2.12. Choose the observation

(2.146) $$z(t) = \left.\frac{\partial}{\partial n}\left(\frac{\partial y}{\partial t}\right)\right|_{\Gamma}$$

which is equivalent to (2.140) where

(2.147) $$\phi \to C\phi = \left.\frac{\partial \phi}{\partial n}\right|_{\Gamma} : D(A) \to H^{1/2}(\Gamma) \subset L^2(\Gamma)$$

is the trace of the normal derivative on Γ.

Example 2.2.13. Choose the observation

(2.148) $$z(t) = \left(\left.\frac{\partial y}{\partial n}\left(\frac{\partial y}{\partial t}\right)\right|_{\Gamma_0}, \left.\frac{\partial y}{\partial t}\right|_{\Gamma_1}\right)$$

which is equivalent to (2.140) with

(2.149) $$\phi \to C\phi = \left(\left.\frac{\partial \phi}{\partial n}\right|_{\Gamma_0}, \left.\phi\right|_{\Gamma_1}\right) : D(A) \to H^{1/2}(\Gamma_0) \times H^{3/2}(\Gamma_1).$$

Example 2.2.14. Choose the observation

$$z(t) = \left(\frac{\partial^4 y}{\partial x^3 \partial t}(0,t), \frac{\partial^3 y}{\partial x^2 \partial t}(0,t), \frac{\partial^2 y}{\partial x\, \partial t}(L,t), \frac{\partial y}{\partial t}(L,t) \right) \tag{2.150}$$

which is equivalent to (2.140) with

$$\phi \to C\phi = \left(\frac{\partial^3 \phi}{\partial x^3}(0), \frac{\partial^2 \phi}{\partial x^2}(0), \frac{\partial \phi}{\partial x}(L), \phi(L) \right) : D(A) \to \mathbb{R}^4. \tag{2.151}$$

2.2.4. Feedback Operators and Smoothness of Solutions. It is extremely interesting to notice that, in all the cases we constructed, the observation operator C is dual to the control operator F

$$C = F^* \quad \text{with } U' \equiv Y \tag{2.152}$$

where the duality is chosen in the appropriate spaces. If a feedback

$$K : Y \to U(\equiv Y') \tag{2.153}$$

is chosen between the observation and control spaces, the closed-loop system will take the form

$$\begin{aligned} &\ddot{y}(t) + Ay(t) + C^* K C \dot{y}(t) = 0, \\ &y(0) = y^0, \qquad \dot{y}(0) = y^1. \end{aligned} \tag{2.154}$$

If K is chosen in such a way that

$$\langle KC\phi, C\phi \rangle_{Y' \times Y} \geq \alpha \|C\phi\|_Y^2 \tag{2.155}$$

for some constant $\alpha > 0$, then the above closed-loop system has a unique solution such that

$$y \in C(0,T;V), \quad \dot{y} \in C(0,T;H) \quad \text{and} \quad C\dot{y} \in L^2(0,T;Y)$$

(cf. Chen, Delfour, Krall, and Payre [18, Thm 2.1]).

2.2.5. Other Types of Damping Operators. We have seen in the previous section that even when the control and observation operators are nastily unbounded, we end up with the usual smoothness on the solution $y(t)$ plus the smoothness we need on the traces of the time derivative of $y(t)$. Some examples of damping and radiation conditions for acoustic and elastic waves can be found in Engquist and Majda [53]. They involve higher-order derivatives.

2.3. Stabilization.

As noted in the Introduction, stability is an important consideration in designing and analyzing control schemes for flexible structures. The concepts of uniform exponential stability and strong stability, which are equivalent for finite-dimensional linear systems, are very different for hyperbolic distributed parameter systems. In this section we will discuss these two stability concepts and describe what is currently achievable for various types of sensor/actuator configurations. In addition, we will review some preliminary stabilization results which obtain even when uniform or strong stability may not. Much of the following material is drawn from Delfour, Lagnese, and Polis [42].

2.3.1. Types of Stability. We take as a starting point the general model (2.37) or (2.38) discussed previously. The norm associated with the state $(y(t), \dot{y}(t))$ in $\mathcal{H} = V \times H$ is

$$\|(y(t), \dot{y}(t))\|_{\mathcal{H}} = (\|y(t)\|_V^2 + \|\dot{y}(t)\|_H^2)^{1/2}, \qquad t \geq 0.$$

The system is then said to be *uniformly exponentially stable* if there exist $M > 0$ and $\beta > 0$ such that for all $t \geq 0$

$$\text{(2.156)} \quad \forall (y^0, y^1) \in V \times H, \qquad \|(y(t), \dot{y}(t))\|_{\mathcal{H}} \leq M e^{-\beta t} \|(y^0, y^1)\|_{\mathcal{H}}.$$

The system is said to be *strongly stable* if

$$\text{(2.157)} \qquad \forall (y^0, y^1) \in V \times H, \qquad \lim_{t \to \infty} \|(y(t), \dot{y}(t))\|_{\mathcal{H}} = 0 \quad \text{in } \mathcal{H}.$$

In engineering applications, active controls are usually finite in number since an infinite number of independent active controllers is difficult to imagine. This means that the set U is finite-dimensional and that the linear continuous control operator $F : U \to V'$ is a finite range operator. Therefore F is a *compact operator*. A continuous linear operator $F : U \to V'$ is said to be compact if it maps bounded subsets U into relatively compact subsets of V'. This is always true when U is finite-dimensional since (1) every bounded subset of a finite-dimensional space is relatively compact and (2) a continuous operator maps compact sets into compact sets. Thus, in order to obtain implementable feedback control strategies it is necessary to incorporate the fact that the control operator is compact.

We first consider the well-known result due to Russell [114] and its extension by Gibson [61] which states that the system (2.37) with $G \equiv 0$ cannot be uniformly exponentially stabilized using compact bounded feedback operators in the sense of (2.94). More precisely, consider the system

$$\text{(2.158)} \qquad \begin{aligned} &\ddot{y}(t) + Ay(t) = K_1 \dot{y}(t) + K_2 y(t) \\ &y(0) = y^0 \in V, \qquad \dot{y}(0) = y^1 \in H \end{aligned}$$

where $K_1 : H \to H$, $K_2 : H \to H$ are continuous linear compact operators, and A a coercive self-adjoint operator $V \to V'$. Russell [114] first proved that system (2.158) could not be uniformly exponentially stable for the case where $K_2 = 0$ and $K_1 = -G$ verifying (2.58) and (2.59). Gibson [61] extended this conclusion to arbitrary continuous and compact linear operators $K_1 : H \to H$ and $K_2 : H \to H$. This Russell–Gibson negative result thus seems to indicate that without inherent damping in the system it is impossible to obtain uniform exponential stabilization using a compact feedback operator. Notice that the resulting feedback operator is bounded from $V \times H$ to $V \times H$ and that $V \times H$ is the state space of the system.

Let us now consider the undamped system with control on the boundary, i.e.,

$$
\begin{aligned}
&\ddot{y}(t) + Ay(t) = Fu(t) \quad \text{in } V' \\
&y(0) = y^0 \in V, \qquad \dot{y}(0) = y^1 \in H.
\end{aligned}
\tag{2.159}
$$

Chen [14]–[17], Lagnese [72], [73], and Chen, Delfour, Krall, and Payre [18] have shown that for this system uniform exponential stabilization can be achieved using boundary feedback. This seems to contradict Gibson's result mentioned in the previous paragraph. To understand what is happening consider a domain $\Omega \subset \mathbb{R}$ in one space dimension. Without loss of generality, we can assume that $\Omega =]0, L[$ and define the boundary

$$\Gamma_0 = \{0\}, \qquad \Gamma_1 = \{1\}.$$

Then in (2.159) $U = \mathbb{R}$ and F is a *continuous linear compact operator with values in* V'. *However, it is not continuous as an operator from* $\mathbb{R}$ *into* $H = L^2(0, L)$ since the evaluation map $v \to v(L)$ is not only not continuous but also not defined for functions v in $L^2(0, L)$. In particular, the hypothesis $F : U \to H$ fails. *This is one of the differences between an "H-continuous" control operator* ($F : U \to H$) *and a "boundary" control operator* ($F : U \to V'$). For bounded connected domains Ω in dimension one, the boundary Γ consists of two points. If the operator A is of order $2m$, there will be at most $2m$ boundary conditions. Thus $U \subset \mathbb{R}^{2m}$ and the control operator will be compact. In conclusion, *Gibson's negative result* [61] *is limited to "H-bounded" control and observation operators and does not apply to "boundary" operators.*

In space dimension two or three the situation is more complex since the boundary Γ is a curve in $\mathbb{R}^2$ or a surface in $\mathbb{R}^3$ which consists of a continuum of points. Thus, in general, the control operator will not be compact. To circumvent this difficulty we must choose control operators of the form

$$
\left\{
\begin{aligned}
&F : \mathbb{R}^N \to V', \\
&\langle F\vec{u}(t), v\rangle_V = \sum_{n=1}^{N} \langle f_n, v\rangle_V \, u_n(t), \\
&\vec{u}(t) = (u_1(t), \cdots, u_N(t)), \qquad \{f_1, \cdots, f_N\} \subset V',
\end{aligned}
\right.
\tag{2.160}
$$

or

$$(2.161) \quad \begin{cases} F : \mathbb{R}^N \to D(A^*)', \\ \langle F\vec{u}(t), v\rangle_{D(A^*)} = \sum_{n=1}^{N} \langle f_n, v\rangle_{D(A^*)} u_n(t), \\ \vec{u}(t) = (u_1(t), \cdots, u_N(t)), \qquad \{f_1, \cdots, f_N\} \subset D(A^*)', \end{cases}$$

depending on the order of the operator A and the type of boundary controls.

Russell's [114] and Gibson's [61] result along with the results we have just mentioned regarding uniform exponential stabilization via boundary feedback raises numerous interesting questions. In particular, we concentrate on the following:

(i) What can be said regarding uniform exponential stabilization when there is inherent damping of various types in the system?

(ii) What can be said regarding uniform exponential stabilization and/or strong stabilization when no damping ($G = 0$) is assumed?

(iii) Since all real sensors and actuators (S/A) average over some (small) region of the spatial domain how realistic is "boundary" feedback or feedback employing point S/A?

(iv) Can anything be said regarding stabilization when even strong stability cannot be established?

(v) How realistic are some of the currently proposed stabilization schemes?

We will now discuss what is currently achievable and what techniques are available for providing some answers to the above questions.

2.3.2. Finite-Dimensional Output Feedback. We first consider systems with damping operator G such that the spectrum is discrete and only a finite number of eigenvalues of the *open-loop system* lie to the right of any vertical line in the complex plane. We further assume that the spectrum is sectorial. Roughly speaking, the spectrum is said to be sectorial if the eigenvalues are contained in a sector as shown in Fig. 2.5. Note that the following damping models are in this category

$$(2.162) \qquad G = 2\gamma A^{1/2} \text{ (Chen–Russell)} \quad \text{with } \sigma_0 = 0,$$

$$(2.163) \qquad G = 2\gamma A \text{ (Voigt–Kelvin)} \quad \text{with } \sigma_0 > 0$$

and generate holomorphic semigroups (cf. Chen and Russell [19]). However, $G = 2\gamma I$ does not.

Under the above assumptions, the problem becomes that of choosing a compensator such that the poles of the closed-loop system all lie to the left

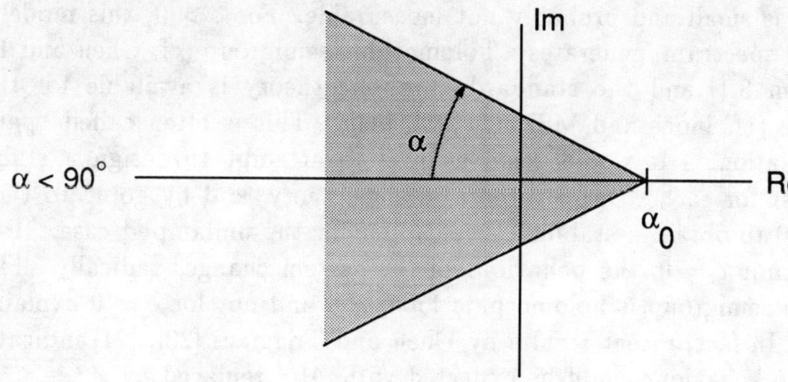

FIG. 2.5. *A sectorial spectrum.*

of some desired vertical line. A number of authors including Curtain [26]–[29], Gibson [61], Sakawa [118], [119], Mousa, Miller, Michel [105] and Nambu [106], [107], have considered various versions of this problem and have derived stabilization schemes for various distributed or boundary control operators. The key point is that the spectrum is assumed to be sectorial, i.e., $\alpha < 90°$. Under this assumption we believe that the infinite-dimensional system (2.12) can be approximated *at the outset* by an Nth order finite-dimensional system. Thus the model becomes that of (2.8) and the stabilization problem is reduced to finding a compensator for a, perhaps very large, finite-dimensional system. A whole body of literature (cf. Balas [3]) exists which treats this problem. An important open question is how large a model is required, i.e., how to choose N. It is our belief that the choice of N can be related to the angle α. We, however, are not aware of any precise results treating this relationship and usually N is chosen somewhat arbitrarily.

At this point it could be argued that the spectrum will always be sectorial and that there is no need to continue with the PDE system model. Although this may be true, there are a number of very good reasons for continuing to work with the PDE model. First, there are physically observed phenomena, such as shocks, which may be completely masked by finite-dimensional models. Second, for lightly damped structures $(\alpha, \gamma) \to (90°, 0)$ the order of the model could be very large and there are numerous open questions, e.g., spillover using noncolocated sensors and actuators, which are difficult and may be obviated by considering the distributed system. Furthermore, the PDE system is in some sense the limiting case and important insight can be gained by considering such cases.

2.3.3. Lightly Damped Structures. As an example of a lightly damped system consider the following model:

$$\ddot{y}(t) + \epsilon A^{1/2}\dot{y}(t) + Ay(t) = Fu(t) \quad \text{in } V',$$
$$y(0) = y^0 \in V, \qquad \dot{y}(0) = y^1 \in H \tag{2.164}$$

where ϵ is small and probably not measurable. For $\epsilon > 0$, this model has a sectorial spectrum, generates a holomorphic semigroup (cf. Chen and Russell [19, Thm 3.1] and the standard parabolic theory is available for treating this case (cf. Lions and Magenes [93], [94]). This is often called "parabolic regularization." It would seem natural to attempt to design a stabilizing controller for each $\epsilon > 0$ by the available theory, and by going to the limit as $\epsilon \to 0$ to obtain a stabilizing controller for the undamped case. However, at the limit $\epsilon = 0$, the behaviour of the system changes radically. The fact that the semigroup is holomorphic for $\epsilon > 0$ and not for $\epsilon = 0$ explains this change. In fact, recent results by Chen and Triggiani [20], [21] indicate that the same behaviour could be expected with $A^{1/2}$ replaced by $A^{\alpha}, \frac{1}{2} \leq \alpha \leq 1$. In the same spirit, Gibson and Rosen [63] showed that his numerical schemes for solving Riccati equations converge for any $\epsilon > 0$ and diverge for $\epsilon = 0$. Such drastic changes indicate that the properties of the system and hence any associated design is highly nonrobust. Thus the lightly damped model (2.164) does not seem to offer any particular advantage over the undamped case. Thus we now turn to results which are available for hyperbolic distributed parameter systems without inherent damping.

2.3.4. Stabilization of Undamped Hyperbolic Systems. The undamped hyperbolic systems that have received the most attention are the wave equation and the beam and plate equations. Exponential stability has been shown to occur when boundary feedback operators are used, for example, in the case

$$\ddot{y}(t) + Ay(t) + K\dot{y}(t) = 0 \tag{2.165}$$

where K is an appropriate linear operator from a function space on the boundary Γ (or part of the boundary $\Gamma_0 \subset \Gamma$) of Ω into another function space on the same piece of the boundary. This result does not seem to depend on the space dimension. Such operators K are usually not compact except in dimension one where the boundary consists of two points. It is important to emphasize that such examples do not fall in the class of compact bounded feedback operators where the results of Russell [114] and Gibson [61] apply.

In fact, in Chen [14]–[17], Lagnese [72], [73], and Chen, Delfour, Krall, and Payre [18], boundary control and observation are used to achieve uniform exponential stability. Can the same result be achieved when one of the two is bounded? If the answer is negative, this would indicate that the negative result on the lack of exponential stability applies to a larger class of systems than the bounded compact linear feedback operators. In space dimension strictly greater than one, we know of no example of exponential stability being achieved using bounded feedback operators with or without compactness. At best, only strong stability is proved.

An important recent development which took place in the analysis of hyperbolic systems is the return to the study of exact controllability and its relationship with the geometry of the space domain. Recently very significant developments have taken place (Lions [91], [92], Leugering [86], [87], Schmidt and Leugering [120]).

A class of control and observation operators that has not been adequately studied from the stabilization point of view is the one where point control and observation or line control and observation (in space dimension two) or surface control and observation (in space dimension three) both occur in the interior of the space domain. Feedback constructions can be made. Moreover, when the resulting feedback operator is symmetric and positive the closed-loop system can be shown to be well posed and dissipative (cf. Chen, Delfour, Krall, and Payre [18]). In fact this result combined with some other results from the theory of dynamical systems would at least indicate that asymptotic stability is achieved in such cases (Dafermos [30]–[32]).

2.3.5. Robustness of the Stabilization Schemes. The stabilization results reviewed in §2.3.4 provide techniques for stabilizing a rather large class of hyperbolic distributed parameter systems. In this section we address the robustness of these results in the face of certain engineering considerations which would arise in implementing the controllers on real systems. In particular, we discuss the robustness with respect to small delays, nonideal sensors and/or actuators, and the addition of small masses on flexible structures. As will be seen, there are no complete answers to any of these questions. Unfortunately, however, there are indications that most of the results may be highly nonrobust.

We first turn to the effect of time delays in boundary feedback stabilization schemes. Most of the following material is taken from Datko, Lagnese, and Polis [39]. Consider the following system:

$$\frac{\partial^2 u}{\partial t^2} - \frac{\partial^2 u}{\partial x^2} + 2a\frac{\partial u}{\partial t} + a^2 u = 0, \qquad 0 < x < 1, \quad t > 0 \tag{2.166}$$

with boundary conditions

$$u(0,t) = 0, \qquad t > 0, \tag{2.167}$$

$$\frac{\partial u}{\partial x}(1,t) = -k\frac{\partial u}{\partial t}(1,t), \qquad t > 0 \tag{2.168}$$

where $a \geq 0$, $k \geq 0$. It is known that (2.166)–(2.167) is uniformly asymptotically stable as long as $a^2 + k^2 > 0$ (see, e.g., Chen [14]). Datko [38], however, has shown that the system will be destabilized by a time delay in the velocity term. Now assume that the boundary condition (2.168) is replaced by

$$\frac{\partial u}{\partial x}(1,t) = -k\frac{\partial u}{\partial t}(1,t-\epsilon), \qquad t > \epsilon \tag{2.169}$$

where $\epsilon > 0$.

THEOREM 2.3.1. (*R. Datko, J. Lagnese, and M. P. Polis* [39]). *Let* $K = e^{-2a}$. *The system* (2.166), (2.167), (2.169) *has the following stability properties:*

(i) *If* $0 < k < (1-K)/(1+K)$, *for each* $\epsilon > 0$ *there exists* $\beta(\epsilon) > 0$ *such that the spectrum of the system lies in* $\mathrm{Re}\,\omega \leq -\beta$.

(ii) *If* $k = (1-K)/(1+K)$, *for each* $\epsilon > 0$ *the spectrum lies in* $\mathrm{Re}\,\omega > 0$, *but there is a countably dense set* R *in* $(0,\infty)$ *such that for each* ϵ *in* R *there is a sequence* $\{\omega_n\}$ *in the spectrum such that*

$$\lim_{n\to\infty} \mathrm{Re}\,[\omega_n] = 0.$$

(iii) *If* $k > (1-K)/(1+K)$, *there is a dense open set* D *in* $(0,\infty)$ *such that for each* ϵ *in* D *the system admits exponentially unstable solutions.*

This example shows that "almost" arbitrary delays can destabilize otherwise stable boundary feedback schemes for both damped and undamped wave equations. This suggests that small delays which will probably occur in computing feedback controls may destabilize certain boundary stabilization schemes—making them nonrobust.

If the delays are known a priori then Skar and Linder [127] have shown that an observer can be used to compensate for the delay. Furthermore, Baras [4] has suggested that frequency domain techniques can be used to derive compensators which are robust to delays. Thus, the question of robustness to delays remains open; however, as indicated by the above example, small delays cannot be simply ignored since they may dramatically affect the stability results.

We now turn to the question of robustness with respect to nonideal sensors and/or actuators. Most of the material here is taken from Delfour, Lagnese, and Polis [42]. All real sensors will yield measurements of the variables being monitored which are weighted averages over a portion of the spatial domain. In the same way the control actions resulting from real actuators affect a distributed part of the spatial domain. Thus, rather than point S/A, a more realistic model would be to consider "concentrated" S/A, whose observations and actions are really distributed but concentrated around a specific spatial location, i.e., weighted averages over small regions of the spatial domain. To clarify the concept of concentrated S/A, consider the one-dimensional problem (2.24) and assume that we want to measure the velocity at a point on the clamped-free beam. The point measurement would be

$$z(t) = \int_0^L \delta(x - x_i)\frac{\partial y}{\partial t}(x,t)\,dx \tag{2.170}$$

where $\delta(x - x_i)$ is the Dirac measure supported at x_i. A concentrated measurement could be modeled as

(2.171) $$z(t) = \int_0^L \xi_i(x) \frac{\partial y}{\partial t}(x,t)dx$$

where ξ_i is some $L^2(0,L)$ function concentrated around x_i. For example, by choosing

(2.172) $$\xi_i(x) = \xi_\rho(x) = \begin{cases} 0, & 0 \le x < L(1-\rho) \\ \frac{1}{L\rho}, & L(1-\rho) \le x \le L \end{cases}$$

where $0 < \rho \le 1$ and

(2.173) $$\int_0^L \xi_\rho(x)dx = 1$$

then (2.171) would yield a measurement of the velocity concentrated around the free end. Furthermore, as $\rho \to 0$ we have $\xi\rho \to \delta(x-L)$ in the sense of distributions.

Now we consider what happens to the stability properties of hyperbolic distributed parameter systems when concentrated S/A are introduced. It is known (cf. [18]) that the one-dimensional clamped beam can be uniformly exponentially stabilized using boundary feedback. However, Delfour, Lagnese, and Polis [42, Thm. 1] introduced concentrated S/A in such a way that the problem falls in the class considered by Russell [114] and Gibson [61], and uniform exponential stability cannot continue to hold. On the other hand, they show that for a rather general class of one-dimensional hyperbolic systems the closed-loop system will be strongly stable using N pairs of collocated S/A if the weightings are chosen in a very special way. Roughly speaking, the necessary and sufficient condition on the weights is that at least one of them not be orthogonal to the span of the eigenfunctions of the uncontrolled system. Unfortunately, the requirement of strong stability places very severe restrictions on the admissible weights and arbitrarily small changes in the weights can lead to destabilization of the system. In reality, the weights will not be known exactly and the admissibility condition is not robust to perturbations in the weights.

We now turn to robustness with respect to the addition of small masses on flexible structures. This is clearly of interest since the S/A masses may be significant. Once again considering the clamped beam which is known to be uniformly exponentially stable using simple boundary feedback schemes, there is numerical evidence (cf. Rideau [111], Delfour, Payre, and Rideau [43], [44]) that adding a mass on the free end of the clamped-free beam (2.24)–(2.26) destroys this exponential stability. Recent theoretical results by Littman and Markus [96] confirm these numerical results.

Based on the above remarks regarding nonrobustness, it would seem that the situation is rather bleak. This is not really the case. First, the rather negative conclusions that can be drawn are all based on a very limited number of examples using simple feedback schemes. Second, the requirement of uniform exponential stability is quite strong for hyperbolic distributed systems and even in this case a reasonably complete theory is beginning to emerge (cf. Lions [89], [90] and Lagnese and Lions [75]). This theory, even if it proves to be nonrobust, is a significant development in our understanding of hyperbolic systems. Furthermore, an idealized theory can be used to provide direction and act as a benchmark for what might be attainable when nonideal cases are being considered. Perhaps most importantly, the nonrobustness may be inherent in the desire to achieve uniform exponential or strong stability results and a weaker stability requirement may be entirely acceptable. The idea here is to consider some form of stability which at least asymptotically assures that the residual energy remains small; which in some sense is "close" to the ideal theory when, for example, concentrated sensors are close to point sensors and time delays are small; and which does not necessitate drastic changes in the form of the feedback compared with the ideal case. Fortunately, some preliminary results in this direction exist, as we shall discuss in the next section.

2.3.6. An Example of Robust Stabilization Using Concentrated S/A. This material is a summary of the main result of Delfour, Lagnese, and Polis [42]. Consider the clamped-free beam controlled by means of collocated S/A concentrated at the free end. The boundary value problem is

$$\ddot{u} + a^4 D^4 u + k^2 \xi \int_0^L \xi \dot{u}\, dx = 0, \qquad 0 < x < L, \quad t > 0 \tag{2.174}$$

$$u(0,t) = Du(0,t) = D^2 u(L,t) = D^3 u(L,t) = 0 \tag{2.175}$$

and where $\xi = \xi_\rho$ is an arbitrary $L^2(0,L)$ function satisfying

$$\xi_\rho \geq 0 \text{ a.e.}, \quad \text{support } \xi_\rho \subset [L(1-\rho), L], \quad \int_0^L \xi_\rho(x)\, dx = 1. \tag{2.176}$$

THEOREM 2.3.2. (*M. C. Delfour, J. Lagnese, M. P. Polis* [42]). *Assume* $E(\dot{u},0) < \infty$ *and that* $\xi_\rho \in L^2(0,L)$ *satisfies* (2.176). *Then for t sufficiently large*

$$E(u,t) \leq \frac{C}{t} E(u,0) + \frac{8}{3}\left(\frac{L}{a}\right)^4 \left(1 - (1-\rho)^{3/2}\right) E(\dot{u},0) \tag{2.177}$$

where C is a constant independent of ρ *and t.*

This theorem shows that whether or not (2.174), (2.175) is strongly stable the energy of the system is bounded by two parts: one that tends uniformly to zero as $t \to \infty$ and a residual that depends only on the initial energy $E(u,0)$ and the support of ξ_ρ. If the support reduces to the point $x = L$, the second part is zero. The remaining inequality is true for smooth initial conditions. By a density argument it is also true for all initial conditions and Pazy's theorem applies to conclude that the energy decays exponentially in agreement with

Chen, Delfour, Krall, and Payre [18]. Thus for the Euler–Bernoulli beam, if the support of the weighting is small and concentrated at the free end the asymptotic energy will also be small. This may be entirely acceptable from an engineering perspective.

It should be noted that Theorem 2.3.2 holds only for the case of collocated S/A that have the same weightings and smooth initial conditions. The case of noncollocated S/A, which is equivalent to assuming different weightings for the sensors than those assumed for the actuators, is still open.

2.4. Open Problems and Promising Directions.

2.4.1. Time-Varying Output Feedback and Moving Controls. In this paper we have emphasized time-invariant control, observation, and feedback operators. However, there are obvious indications that the time dimension might very well play an important role in the stabilization of structures and that the observed lack of robustness we have encountered so far might be the implicit price to pay for neglecting the fundamental interaction of time and space in hyperbolic systems. It is very unlikely that an abstract theory of time-varying feedback control of hyperbolic systems can be developed without explicitly taking into account the geometry of the domain.

One class of control which falls into this category are the moving or movable controls (cf., for instance, Butkovsky [10], Butkovsky, Darinskiy, and Pustilnikov [11], Martin [101], Butkovsky and Pustilnikov [13], McLaughlin and Slemrod [103]). Some serious mathematical work also exists on the localization of the support of the actuators for linear parabolic partial differential equations (Zolésio [139]). This raises many engineering and mathematical questions such as the choice of time-varying observers (cf. Baras and Bensoussan [5]). If we accept the fact that hyperbolic systems can be controlled and observed by only a finite number of actuators and sensors, the "time" ingredient would seem to be a good way to restore the necessary infinite-dimensionality to the control action.

2.4.2. Geometric Control (Shape and/or Structure). In structures, as opposed to heat problems, the shape of the domain or the configuration of the structure could also be used as control variables. So the geometry, which is a critical variable in establishing exact controllability, can be used to stabilize the structure. This type of problem has been studied by a limited number of authors for the wave equation (Cooper and Bardos [24], Bardos and Chen [6], Zolésio and Truchi [141], Truchi [137], Truchi and Zolésio [138]) and for the plate equation (Zolésio [140]). In Zolésio [140] a periodic control law is constructed in order to "pump out" the transversal vibration energy. Numerical tests for the wave equation indicate that for each initial condition the decay is at least exponential.

In our opinion, this is a most promising area of development for the future. It involves ideas and techniques from the theory of hyperbolic partial differential equations with time varying domains and from the fundamental theory of shape optimal design.

2.4.3. Frequency Domain Techniques. In this paper we have completely ignored frequency domain techniques or techniques based on a transformation of the time variable. This point of view is intrinsically interesting and in a certain way vital for structural problems. The reader will find a good account of this type of technique in Francis, Helton, and Zames [58]. Such techniques have also been used extensively in the structural engineering literature (cf. Piché [108]). For the recent use of H^∞-theory for infinite-dimensional systems, the reader is also referred to Flamm [55] and Foias, Tannenbaum, and Zames [56], [57].

2.4.4. Dissipation and Nonlinearity. Stabilization of flexible structures has generally been studied with linear or linearized models. Nonlinear models with their obviously more complex structural dynamics have been slower to appear in the control literature (cf. Baillieul and Levi [1]), Krishnaprasad and Marsden [70]. It seems that nonlinear models contain inherent damping mechanisms which may be completely lost in the linearization process.

2.4.5. Pointing Accuracy, Nonlinear Elasticity. When compared to robots operating, for example, in the atmosphere, large flexible structures are not only large but also fragile and very lightly damped. Reasonable hypotheses used in the theory of small, bulky, and heavily damped structures are no longer reasonable. So we are facing a fundamentally different new issue.

Looking at the vibration of a short bulky beam, the deflection will be small and good accuracy can be achieved in controlling the tip of the beam. For a long thin beam, the tip deflection will be large and accuracy at the tip of the beam will be poor. In both cases, the linear elasticity theory will be applicable and correct. However, "small" is a relative concept which refers to the deflection as compared to the length of the bar. Unfortunately, pointing accuracy does not follow this hypothesis. So it is very likely that for a given required accuracy the linear elasticity might not be sufficient to fulfill this task.

In addition, the sensors and actuators will probably have dead zones which, of course, introduce other nonlinearities into the equations. For large slewing maneuvers the linearization of the vehicle dynamics will no longer be acceptable. How to model and treat these remains a complex question. Techniques such as those suggested by Baillieul and Levi [1], Krishnaprasad [69], Krishnaprasad and Marsden [70], Simo, Marsden, and Krishnaprasad [126] are certainly a significant step to better capture the real behaviour of the structure.

2.5. Conclusions.

In this paper we have given a general control model which encompasses most linear time-invariant finite-dimensional or partial differential equation models often used to describe large flexible space structures. The model is sufficiently general that the various memoryless damping models available for flexible structures can be easily incorporated as special cases. In addition, the model can be extended to account for damping with memory. Furthermore, various types of sensor/actuator (S/A) models can be readily introduced into the model. The importance of the model lies in its generality and the possibility, simply by associating the correct spaces with the various operators, of incorporating a wide variety of damping and S/A models.

Three types of S/A models and the operators associated with them have been discussed, i.e., point, boundary, and distributed models. It has been noted that in order to derive implementable feedback control strategies the control and observation operators should be of finite range. If, in addition, linear feedback is assumed between observation and control then we know the following:

- If the control and observation operators are bounded and if the operator FKC is compact, Russell [114] and Gibson [61] have shown that exponential stability cannot be obtained unless the open-loop system itself is exponentially stable;

- If one of the two operators is bounded and the other is a point or boundary operator, then we only know of examples where strong stability can be achieved without inherent damping;

- If both operators are point or boundary operators, then exponential stability results can be obtained for undamped systems.

Clearly, inherent damping plays a key role in the robustness of existing stability results. For undamped systems, the stability results seem highly nonrobust to small delays, small changes in the position of the S/A or small changes in the support of the S/A. An important open question is the precise way in which the inherent damping and the robustness of the closed-loop system are related. This nonrobustness would seem to indicate that richer control or stabilizing mechanisms such as time-varying feedback employing moving controls or geometric control should be pursued.

In closing, we note that even for the worst case, i.e., undamped models, very impressive exponential stabilization results using boundary feedback are available and these results can serve as a benchmark (as was done by Delfour, Lagnese, and Polis [42]) in obtaining more robust stability results when some of the idealized S/A models are relaxed.

Acknowledgment.

We thank the reviewer for helpful comments and suggestions. This research has been supported in part by Department of Communications (Ottawa, Canada) contracts DSS.24ST.36001-3-1398, 20ST.36100-4-4120, 12ST.36100-5-0110, 41ST.36100-6-4156 and Natural Sciences and Engineering Research Council of Canada, grants OGP 0008730, INF 0007939 and OGP 0008670.

References

[1] J. Baillieul and M. Levi, *Dynamics of Rotating Flexible Structures*, in Proc. 22nd IEEE Conference on Decision and Control, IEEE Publications, Dec. 1983, pp. 808–813.

[2] A. V. Balakrishnan, *A theory of nonlinear damping in flexible structures*, in Boundary Control and Boundary Variations, J. P. Zolésio, ed., Springer-Verlag, Berlin, New York, 1988, pp. 18–37.

[3] M. J. Balas, *Trends in large space structure control theory: Fondest hopes, wildest dreams*, IEEE Trans. Automatic Control, AC-27 (1982), pp. 522–535.

[4] J. S. Baras, Private communication.

[5] J. S. Baras and A. Bensoussan, *On observer problems for systems governed by partial differential equations*, Systems Research Center, Report SRC-TR-86-47, University of Maryland, College Park, MD, 1986.

[6] C. Bardos and G. Chen, *Control and stabilization for the wave equation, Part* III: *domain with moving boundary*, SIAM J. Control Optim., 19 (1981), pp. 123–138.

[7] B. M. Barry, *The Minimum Stress Problem for Nonlinear Hybrid Train Models*, Ph.D. thesis, Queen's University, Kingston, Canada, 1982.

[8] S. K. Biswas and N. U. Ahmed, *Stabilization of a class of hybrid systems arising in flexible spacecraft*, J. Optim. Theory Appl., 50 (1986), pp. 83-108.

[9] L. Boltzmann, *Zur Theorie elastischen Nachwirkung*, Ann. Phys., 7 (1876), Erganzugsband, pp. 624–625.

[10] A. G. Butkovsky, *Some new results in distributed parameter system control*, in Proc. Third Sympos. on Control of Distributed Parameter Systems, J. P. Barbary and L. Le Letty, eds., IFAC Publications, Pergamon Press, Toulouse, France, 1982, pp. SP 29–44.

[11] A. G. Butkovsky, Yu V. Darinskiy, and L. M. Pustilnikov, *Control of distributed systems by displacement of the source*, Automat. i Telemeh., 5 (1974); Automat. Remote Control, 35 (1974), pp. 701–719. (In English.)

[12] ——, *Stationary control with systems having distributed parameter*, Automat. i Telemeh., 2 (1976), pp. 15–25. (In Russian.) Automat. Remote Control, 37 (1976), pp. 133–142. (In English.)

[13] A. G. Butkovsky and L. M. Pustilnikov, *Theory of movable control*, Nauka, Moscow, 1980.

[14] G. Chen, *Energy decay rates and exact boundary value controllability for the wave equation in a bounded domain*, J. Math. Pures Appl., 58 (1979), pp. 249–274.

[15] ——, *A note on the boundary stabilization of the wave equation*, SIAM J. Control Optim., 19 (1981), pp. 106–113.

[16] ——, *Control and stabilization for the wave equation in a bounded domain*, SIAM J. Control Optim., 17 (1979), pp. 66–81.

[17] ——, *Control and stabilization for the wave equation in a bounded domain, Part* II, SIAM J. Control Optim., 19 (1981), pp. 114–122.

[18] G. Chen, M. C. Delfour, A. Krall, and G. Payre, *Modeling, stabilization and control of serially connected beams*, SIAM J. Control Optim., 25 (1987), pp. 526–546.
[19] G. Chen and D. L. Russell, *A mathematical model for linear elastic systems with structural damping*, Quart. Appl. Math., 39 (1982), pp. 433–454.
[20] S. Chen and R. Triggiani, *Proof of two conjectures by G. Chen and D. L. Russell on structural damping for elastic systems*, in Proc. of the Seminar in Approximation and Optimization (Havana, Cuba, January 12–14, 1987), pp. 234–256, Springer-Verlag, Berlin, New York, 1987.
[21] ——, *Proof of extension of two conjectures on structural damping for elastic systems: the case* $1/2 < \alpha < 1$, Pacific J. Math., 136 (1989), pp. 15–55.
[22] R. W. Clough and J. Penzien, *Dynamics of structures*, McGraw-Hill, New York, 1975.
[23] J. Cooper, *Local decay of solutions of the wave equation in the exterior of a moving body*, J. Math. Anal. Appl., 49 (1975), pp. 130–153.
[24] J. Cooper and C. Bardos, *A nonlinear wave equation in a time-dependent domain*, J. Math. Anal. Appl., 42 (1973), pp. 29–60.
[25] J. Cooper and W. A. Strauss, *Energy boundedness and decay of waves reflecting off a moving obstacle*, Indiana Univ. Math. J., 25 (1976), pp. 671–690.
[26] R. F. Curtain, *Finite dimensional compensators for some hyperbolic systems with boundary control*, in Control Theory for Distributed Parameter Systems and Applications, pp. 77–91, Lecture Notes Control Inform. Sci., 54, Springer-Verlag, Berlin, 1983.
[27] ——, *Observation and boundary control via integral dynamic output feedback of a finite dimensional compensator*, Report TW-243, Mathematisch Instituut, Rijksniversiteit, Groningen, the Netherlands, August 1982.
[28] ——, *Pole assignment for distributed systems by finite dimensional control*, Automatica, 21 (1985), pp. 57–67.
[29] ——, *Compensators for infinite dimensional linear systems: a survey*, J. Franklin Institute, 315 (1983), pp. 331–346.
[30] C. M. Dafermos, *Uniform processes and semicontinuous Lyapunoff functionals*, J. Differential Equations, 4 (1968), pp. 57–65.
[31] ——, *Wave equations with weak damping*, SIAM J. Appl. Math., 18 (1970), pp. 759–767.
[32] ——, *An invariance principle for compact processes*, J. Differential Equations, 9 (1971), pp. 239–252.
[33] ——, *Applications of the invariance principle for compact processes. I. Asymptotically dynamical systems*, J. Differential Equations, 9 (1971), pp. 291–299. II. *Asymptotic behavior of solutions of a hyperbolic conservative law*, J. Differential Equations, 11 (1972), pp. 416–424.
[34] ——, *Uniform processes and semicontinuous Liapunov functionals*, J. Differential Equations, 11 (1972), pp. 401–415.
[35] ——, *Asymptotic stability in viscoelasticity*, Arch. Rational Mech. Anal., 37 (1970), pp. 297–308.
[36] ——, *The entropy rate admissibility criterion for solutions of hyperbolic conservation laws*, J. Differential Equations, 14 (1973), pp. 202–212.
[37] ——, *The Riemann problem for certain classes of hyperbolic systems of conservation laws*, J. Differential Equations, 20 (1976), pp. 90–114.
[38] R. Datko, *Representation of solutions and stability of linear differential-difference equations in a Banach space*, J. Differential Equations, 29 (1978), pp. 105–166.

[39] R. Datko, J. Lagnese, and M. P. Polis, *An example on the effect of time delays in boundary feedback stabilization of wave equations*, SIAM J. Control Optim., 24 (1986), pp. 152–156.

[40] J. H. Davis and B. M. Barry, *A distributed model for stress control in multiple locomotive trains*, J. Appl. Math. Optim., 3 (1977), pp. 163–190.

[41] J. de Lafontaine and M. E. Steiber, *Sensor/actuator selection and placement for control and elastic continuum*, in Control of Distributed Parameter Systems, Proc. Fourth IFAC Sympos. on Control of Distributed Parameters, Los Angeles, CA, June 30–July 2, 1986.

[42] M. C. Delfour, J. Lagnese, and M. P. Polis, *Stabilization of hyperbolic systems using concentrated sensors and actuators*, IEEE Trans. on Automatic Control, AC-31 (1986), pp. 1091–1096.

[43] M. C. Delfour, G. Payre, and P. Rideau, *Calcul des valeurs propres pour des structures linéaires par la méthode de Kuhn*, in Analysis and Optimization of Systems, A. Bensoussan and J. L. Lions, eds., Springer-Verlag, Berlin, New York, 1986, pp. 114–128.

[44] ——, *Numerical computation of eigenvalues for linear structures by Kuhn's method*, in Control of Distributed Parameter Systems, H. E. Rauch, ed., Pergamon Press, Oxford, New York, 1986, pp. 147–153.

[45] M. C. Delfour and J. P. Zolésio, *Shape sensitivity analysis via Min Max differentiability*, SIAM J. Control Optim., 26 (1988), pp. 834–862.

[46] ——, *Shape sensitivity analysis via penalization*, Ann. Mat. Pura Appl., 151 (1988), pp. 179–212.

[47] ——, *Analyse des problèmes de forme par la dérivation des Min Max*, in Analyse Non Linéaire, H. Attouch, J. P. Aubin, F. H. Clarke, and I. Ekeland, eds., Analyse Non Linéaire, Ann. l'Institut Henri-Poincaré, Special volume in honor of J.-J. Moreau, Gauthier-Villars, Bordas, Paris, 1989, pp. 211–228.

[48] ——, *Anatomy of the shape Hessian*, Ann. Mat. Pura Appl., 159 (1991), pp. 315–339.

[49] ——, *Shape analysis and optimization in distributed parameter systems*, in Control of Distributed Parameter Systems, M. Amouroux and A. El Jai, eds., IFAC Symposia series 3, Pergamon Press, Oxford, New York, 1990, pp. 35–43.

[50] ——, *Computation of the shape Hessian by a Lagrangian method*, in Control of Distributed Parameter Systems, M. Amouroux and A. El Jai, eds., IFAC Symposia series 3, Pergamon Press, Oxford, New York, 1990, pp. 215–220.

[51] ——, *Shape optimization and analysis problems: Lagrangian methods and applications*, in Control of Boundaries and Stabilization, J. Simon, ed., Berlin, New York, 1989, pp. 95–108.

[52] ——, *Velocity method and Lagrangian formulation for the computation of the shape Hessian*, SIAM J. Control Optim., 29 (1991), pp. 1414–1442.

[53] B. Engquist and A. Majda, *Radiation boundary conditions for acoustic and elastic wave calculations*, Comm. Pure Appl. Math., 32 (1979), pp. 313–357.

[54] H. O. Fattorini, *Boundary control systems*, SIAM J. Control, 6 (1968), pp. 349–385.

[55] D. S. Flamm, *Control of Delay Systems for MiniMax Sensitivity*, Ph.D. thesis, Massachusetts Institute of Technology, Cambridge, MA, 1986.

[56] C. Foias, A. Tannenbaum, and G. Zames, *Weighted sensitivity minimization for delay systems*, IEEE Trans. Automatic Control, AC-31 (1986), pp. 763–766.

[57] ——, *On the H^∞-optimal sensitivity problem for systems with delays*, SIAM J. Control Optim., 25 (1987), pp. 686–705.

[58] B. A. Francis, J. W. Helton, and G. Zames, *H^∞-optimal feedback controllers for linear multivariable systems*, IEEE Trans. Automatic Control, AC-29 (1984), pp. 888–900.

[59] N. Fujii, *Feedback stabilization of distributed parameter systems by a functional observer*, SIAM J. Control, 18 (1980), pp. 108–120.

[60] T. Geveci, *Boundary controllability of hyperbolic partial differential equations*, SIAM J. Control Optim., 19 (1981), pp. 359–367.

[61] J. S. Gibson, *A note on stabilization of infinite dimensional linear oscillators by compact linear feedback*, SIAM J. Control Optim., 18 (1980), pp. 311–316.

[62] ——, *An analysis of optimal modal regulation: convergence and stability*, SIAM J. Control Optim., 19 (1981), pp. 686–707.

[63] J. S. Gibson and I. G. Rosen, *Numerical approximation for the infinite-dimensional discrete-time optimal linear-quadratic regulator problem*, SIAM J. Control Optim., 26 (1988), pp. 428–451.

[64] Lop Fat Ho, *Observabilité frontière de l'équation des ondes*, C. R. Acad. Sci. Paris, Sér. A, 302 (1986), pp. 443–446.

[65] P. C. Hughes, *MSAT structural flexibility and control assessment*, Dynacon Enteprises Ltd., Report MSAT-1, March 1981, Toronto, Canada.

[66] ——, *Modeling of Energy Dissipation (Damping) in Flexible Space Structures*, Department of Communications, University of Ottawa, Canada, Report DOC-CR-SP-82-024, June, 1982.

[67] T. L. Johnson, *Progress in modelling and control of flexible spacecraft*, J. Franklin Institute, 315 (1983), pp. 495–520.

[68] V. Kolousek, *Dynamics in Engineering Structures*, Butterworths, London, 1973.

[69] P. S. Krishnaprasad, *Lee–Poisson structures, dual-spin spacecraft and asymptotic stability*, Nonlinear Anal. T.M.A., 9 (1985), pp. 1011–1035.

[70] P. S. Krishnaprasad and J. E. Marsden, *Hamiltonian structures and stability for rigid bodies with flexible attachment*, Arch. Rational Mech. Anal., 48 (1986), pp. 71–73.

[71] Chau-Chin Kwan and Kang-Ning Wang, *Sur la stabilisation de la vibration élastique*, Scientia Sinica, 17 (1974), pp. 446–467.

[72] J. Lagnese, *Decay of solutions of wave equations in a bounded region with boundary dissipation*, J. Differential Equations, 50 (1983), pp. 163–182.

[73] ——, *Boundary stabilization of linear elastodynamic systems*, SIAM J. Control Optim., (1983), pp. 968–984.

[74] ——, *Stabilization of Plates*, in Proc. 3rd Internat. Conf. on Distributed Systems, Lecture Notes in Control Inform. Sci. 102, Springer-Verlag, Berlin, 1988, pp. 204–215.

[75] J. Lagnese and J. L. Lions, *Modelling, Analysis and Control of Thin Plates*, Masson, Paris, 1988.

[76] J. Lagnese, M. C. Delfour, and M. P. Polis, *Stabilization of hyperbolic systems using concentrated sensors and actuators*, in Proc. 24th Conference on Decision and Control, IEEE Publications, New York, 1985, pp. 2005–2006.

[77] I. Lasiecka, J. L. Lions, and R. Triggiani, *Non homogeneous boundary value problems for second order hyperbolic operators*, J. Math. Pures. Appl., 65 (1986), pp. 149–192.

[78] I. Lasiecka and R. Triggiani, *Hyperbolic equations with Dirichlet boundary feedback via position vector: regularity and almost periodic stabilization, Part* I, J. Appl. Math. Optim., 8 (1981), pp. 1–37; *Part* II, 8 (1982), pp. 103–130; *Part* III, 8 (1982), pp. 199–221.

[79] ——, *Dirichlet boundary stabilization of the wave equation with damping*, in Proc. Third IFAC Sympos. on Control of Distributed Parameter Systems, Toulouse, France, 1982, pp. IV 1–4.

[80] I. Lasiecka and R. Triggiani, *Dirichlet boundary stabilization of the wave equation with damping feedback of finite range*, J. Math. Anal. Appl., 97 (1983), pp. 112–130.

[81] ——, *Regularity of hyperbolic equations under $L^2(0,T;L^2(\Gamma))$-Dirichlet boundary terms*, Appl. Math. Optim., 10 (1983), pp. 275–286.

[82] ——, *Finite rank, relatively bounded perturbations of semigroups generators. Part II: Spectrum and Riesz basis assignment with applications to feedback systems*, Ann. Mat. Pura Appl., 143 (1986), pp. 47–100.

[83] ——, *Feedback semigroups and cosine operators for boundary feedback parabolic and hyperbolic equations*, J. Differential Equations, 47 (1983), pp. 246–272.

[84] ——, *Exponential uniform stabilization of the wave equation with $L_2(0,\infty;L_2(\Gamma))$ boundary feedback acting in the Dirichlet boundary conditions*, in Proc. 24th Conference on Decision and Control, IEEE Publications, New York, 1985, pp. 123–125.

[85] ——, *Riccati equations for hyperbolic partial differential equations with $L_2(0,T;L_2(\Gamma))$ Dirichlet boundary terms*, Proc. 24th Conference on Decision and Control, IEEE Publications, New York, 1985, pp. 114–116.

[86] G. Leugering, *On boundary controllability of viscoelastic systems*, in Optimal Control of Partial Differential Equations, A. Bermudez, ed., Springer-Verlag, Berlin, New York, 1989, pp. 190–201.

[87] ——, *Boundary control of a viscoelastic string*, in Volterra Integrodifferential Equations in Banach Spaces and Applications, G. Da Prato and M. Iannelli, eds., Pitman Res. Notes in Math. Ser., 190, Longman, Essex, England, 1987, pp. 258–270.

[88] N. Levan, *The stabilization problem: a Hilbert space operator decomposition approach*, IEEE Trans. Automatic Control, AC-25 (1978), pp. 721–727.

[89] J. L. Lions, *Contrôle optimal des systémes gouvernés par des équations aux dérivées partielles*, Dunod, Paris, 1968.

[90] ——, *Boundary and pointwise control for distributed systems*, this volume, 1992, pp. 1-39.

[91] ——, *Exact controllability, stabilization and perturbations for distributed systems*, The John Von Neumann Lecture, SIAM National Meeting, Boston, MA, July 21–25, 1986.

[92] ——, *Controllabilité exacte des systèmes distribués remarques sur la théorie générales et les applications*, in Analysis and Optimization of Systems, A. Bensoussan and J. L. Lions, eds., Springer-Verlag, Berlin, New York, 1986, pp. 3–14.

[93] J. L. Lions and E. Magenes, *Problèmes aux limites non homogènes et applications*, Vol. 2, Dunod, Paris, 1968.

[94] ——, *Problèmes aux limites non homogènes et applications*, Vol. 1, Dunod, Paris, 1968.

[95] W. Littman, *Boundary control theory of beams and plates*, in Proc. 24th IEEE Conference on Decision and Control, Fort Lauderdale, FL, 1985, pp. 2007–2009.

[96] W. Littman and L. Markus, *Stabilization of a hybrid system of elasticity by feedback boundary damping*, Ann. Mat. Pura Appl., 152 (1988), pp. 281–330.

[97] D. L. Lukes and D. L. Russell, *The quadratic criterion for distributed systems*, SIAM J. Control, 7 (1969), pp. 101–121.

[98] M. K. Mackay, *Active control of large flexible space structures*, Ph.D. dissertation, U.C.L.A., Los Angeles, CA, 1983.

[99] A. Majda, *Disappearing solutions for the dissipative wave equation*, Indiana Univ. Math. J., 24 (1975), pp. 1119–1133.

[100] ——, *The location of the spectrum for the dissipative acoustic operator*, Indiana Univ. Math. J., 25 (1976), pp. 973–987.

[101] J. C. E. Martin, *Optimal scanning control of parabolic systems*, Internat. J. Control, 21 (1975), pp. 305–317.

[102] ——, *On an optimal scanning control problem in a one-dimensional space*, IEEE Trans. Automatic Control, AC-22 (1977), pp. 667–669.

[103] J. R. McLaughlin and M. Slemrod, *Scanning control of a vibrating string*, Appl. Math. Optim., 14 (1986), pp. 27–47.

[104] L. Meirovitch, *Analytical Methods in Vibrations*, MacMillan, New York, 1967.

[105] M. Mousa, R. K. Miller, and A. N. Michel, *Stability analysis of hybrid composite dynamical systems involving operators and difference equations*, IEEE Trans. Automatic Control, AC-31 (1986), pp. 603–615.

[106] T. Nambu, *Feedback stabilization for distributed parameter systems of parabolic type*, J. Differential Equations, 33 (1979), pp. 167–188.

[107] ——, *Feedback stabilization of diffusion equations by a functional observer*, J. Differential Equations, 43 (1982), pp. 257–280.

[108] R. A. Piché, *Analysis of Structural Control Problems Using Frequency-Domain Continuum Models*, Ph.D. thesis, University of Waterloo, Waterloo, Canada, 1986.

[109] E. P. Popov, *Introduction to Mechanics of Solids*, Prentice-Hall, Englewood Cliffs, NJ, 1968.

[110] J. P. Quinn and D. L. Russell, *Asymptotic stability and energy decay rates for solutions of hyperbolic equations with boundary damping*, Proc. Roy. Soc. Edinburgh Sect. A, 77 (1977), pp. 97–127.

[111] P. Rideau, *Contrôle d'un assemblage de poutres flexibles par des capteurs—actionneurs ponctuels: Étude du spectre du système*, thése de docteur ingénieur, Sophia-Antipolis, France, 1985.

[112] A. P. Ross, *The Stability and Stabilizability of Infinite Dimensional Linear Systems via Liapunov's direct method*, Ph.D. dissertation, U.C.L.A., Los Angeles, CA, 1979.

[113] D. L. Russell, *Linear stabilization of the linear oscillator in Hilbert space*, J. Math. Anal. Appl., 25 (1969), pp. 663–675.

[114] ——, *Decay rates for weakly damped systems in Hilbert space obtained via control theoretic methods*, J. Differential Equations, 19 (1975), pp. 344–370.

[115] ——, *On mathematical models for the elastic beam with frequency-proportional damping*, this volume, (1992), pp. 125–170.

[116] ——, *A unified boundary controllability theory for hyperbolic and parabolic equations*, Stud. Appl. Math., 52 (1973), pp. 189–211.

[117] ——, *Controllability and stabilization theory for linear partial differential equations. Recent progress and open questions*, SIAM Rev., 20 (1978), pp. 639–739.

[118] Y. Sakawa, *Feedback control of the second order evolution equations with unbounded observation*, Second Vorau Conference, July 9–14, 1984.

[119] ——, *Feedback control of second order evolution equations with damping*, SIAM J. Control Optim., 22 (1984), pp. 343–361.

[120] E. J. P. G. Schmidt and G. Leugering, *Boundary control of a viscoelastic plate with internal damping*, Internal Report, McGill University, Montreal, Quebec, Canada, May 1987.

[121] J. M. Schumacher, *A direct approach to compensator design for distributed parameter systems*, SIAM J. Control Optim., 21 (1983), pp. 823–836.

[122] ——, *Dynamic feedback in finite and infinite dimensional linear systems*, Mathematical Centre Tracts, 143, Mathematish Centrum, Amsterdam, 1981.

[123] Shih Chu-Chung, *Sur une classe d'espaces fonctionnels intervenant en contrôle optimal*, C. R. Acad. Sci. Paris, 290 (1980), pp. 761–764.

[124] Shih Chu-Chung and J. Simon, *Sur un espace intervenant en contrôle ponctuel*, C. R. Acad. Sci. Paris, 290 (1980), pp. 693–696.

[125] S. Shunhua, *Boundary stabilization of hyperbolic systems with no dissipative conditions*, SIAM J. Control Optim., 20 (1982), pp. 862–883.

[126] J. C. Simo, J. E. Marsden and P. S. Krishnaprasad, *The Hamiltonian structure of nonlinear elasticity: the convective representation of solids, rods and plates*, Arch. Rational Mech. Anal., 104 (1988), pp. 125–183.

[127] S. Skar and D. W. Linder, *On correcting the destabilizing effects of delay in boundary control*, in Proc. 24th IEEE CDC, Vol. 1, Fort Lauderdale, FL, 1985, pp. 126–127.

[128] E. Skudrzyk, *Simple and Complex Vibratory Systems*, Pennsylvania State University Press, University Park, PA, 1968.

[129] M. Slemrod, *The linear stabilization problem in Hilbert space*, J. Functional Anal., 11 (1972), pp. 334–345.

[130] ——, *A note on complete controllability and stabilizability for linear control systems in Hilbert space*, SIAM J. Control, 12 (1974), pp. 500–508.

[131] ——, *Stabilization of boundary control systems*, J. Differential Equations, 22 (1976), pp. 402–415.

[132] A. E. Taylor, *Introduction to Functional Analysis*, John Wiley, New York, 1958.

[133] R. Triggiani, *On the stabilizability problem in Banach space*, J. Math. Anal. Appl., 52 (1975), pp. 383–403, Addendum to the paper, *On the stabilizability problem in Banach space*, J. Math. Anal. Appl., 56 (1976), pp. 492–493.

[134] ——, *On Nambu's boundary stabilizability problem for diffusion processes*, J. Differential Equations, 33 (1979), pp. 189–200.

[135] ——, *Well-posedness and regularity of boundary feedback parabolic systems*, J. Differential Equations, 36 (1980), pp. 347–362.

[136] ——, *Boundary feedback stabilizability of parabolic equations*, J. Appl. Math. Optim., 6 (1980), pp. 201–220.

[137] C. Truchi, *Stabilisation par variation du domaine*, Thèse de doctorat, Université de Nice, Nice, France, 1987.

[138] C. Truchi and J. P. Zolésio, *Wave equation in time periodical domains*, in Stabilization of Flexible Structures, A. V. Balakrishnan and J. P. Zolésio, eds., Optimization Software, Inc., Publications Division, New York, Los Angeles, 1988, pp. 282–294.

[139] J. P. Zolésio, *Identification de domaines par déformation*, Thèse de doctorat d'état, Université de Nice, Nice, France, 1979.

[140] ——, *Shape stabilization of flexible structures*, in Distributed Parameter Systems, F. Kappel, K. Kunish, and W. Schappacher, eds., Springer-Verlag, Berlin, New York, 1985, pp. 446–460.

[141] J. P. Zolésio and C. Truchi, *Shape stabilization of wave equation*, in Boundary Control and Boundary Variations, J. P. Zolésio, ed., Springer-Verlag, Berlin, New York, 1988, pp. 372–398.

Chapter 3

A Comparison of Three Approximation Schemes for Optimal Control of a Flexible Structure

J. S. Gibson and A. Adamian

3.1. Introduction.

For a distributed model of a flexible structure, the linear-quadratic-Gaussian (LQG) optimal control problem on the infinite time interval leads to an optimal compensator that is infinite-dimensional. Like the finite-dimensional LQG problem, the infinite-dimensional problem separates into a deterministic linear-quadratic regulator problem and a state estimation (Kalman filtering) problem, each of which leads to a Riccati equation whose solution determines the optimal gains. The infinite-dimensional problem differs from the finite-dimensional problem in that the Riccati equations and optimal gains involve infinite-dimensional operators instead of matrices. This means that the solution to the infinite-dimensional problem must be approximated numerically, usually by solving a sequence of finite-dimensional LQG problems whose solutions converge in some sense to the solution of the infinite-dimensional problem. Also, since only finite-dimensional compensators can be realized, the infinite-dimensional compensator must be approximated for implementation.

The approximation theory for numerical solution of the infinite-dimensional LQG problem for flexible structures was established in [5], [6], and numerical results for distributed models of complex structures were given in [6], [10], [13]. Each of the approximating finite-dimensional LQG problems yields a finite-dimensional compensator, which approximates the optimal infinite-dimensional compensator. The convergence of the sequence of finite dimensional compensators to the infinite-dimensional compensator indicates the order of the finite-dimensional compensator that will produce an essentially optimal closed-loop system. Because it is usually impossible to predict a priori what order compensator will be necessary to provide near optimal, or even acceptable, performance, the convergence analysis and criteria for the sequence of implementable compensators are the main practical results of the approximation theory for the infinite LQG problem.

The numerical solution of the LQG problem is based on an approximation of the flexible structure with either finite elements or natural modes.

Although many different approximation schemes for the structure can be used, the approximating control and estimator gains and the finite-dimensional compensators produced by all correct approximation schemes will converge to the same optimal infinite-dimensional gains and compensator under reasonable conditions on the distributed model of the structure. (For example, coercive damping usually is needed in all flexible components.) This paper presents a numerical study of three structural approximation schemes that can be used as a basis for approximate solution of the LQG problem for a large class of flexible structures. For this study, we use a structure that consists of a flexible beam attached at one end to a rotating rigid hub and at the other end to a concentrated mass. In its essential characteristics (flexibility, rigid-body mode, multiple bodies) this structure is similar to more complex aerospace structures, which are the primary motivation for this work.

We should point out that, in general, there are two stages of approximation in designing an implementable near-optimal compensator for a distributed system. The first stage obtains the optimal compensator as the limit of a sequence of finite-dimensional compensators; the second stage reduces, if possible, the order of a large finite-dimensional compensator that is essentially identical to the optimal compensator. The first stage of approximation is the subject of this paper, although the two are not always independent. We will discuss the order of implementable compensators for our example briefly in §3.7.

Two of the approximation schemes use finite element approximations of the beam. The first scheme uses cubic Hermite splines as basis functions, and the second scheme uses cubic B-splines as basis functions. Cubic Hermite splines are the basis functions used most commonly in engineering finite element approximations of beams. Cubic B-splines have continuous second derivative (bending moment) whereas the Hermite splines have discontinuous second derivative at the nodes. For the same number of beam elements, the Hermite splines have twice as many degrees of freedom as the B-splines. Rosen [15] has demonstrated that B-splines are superior to both Hermite splines and component modes for computing the natural modes of vibration for an Euler–Bernoulli beam with a tip body. The third approximation scheme uses as basis functions the natural mode shapes of undamped, free vibration of the structure. To get these mode shapes, we must use a preliminary finite element approximation of the structure. We use the Hermite splines for this.

In the numerical comparison of the three approximation schemes, we first compare how the finite-dimensional compensators produced by each scheme converge to the infinite-dimensional compensator, and then we compare the eigenvalues of the closed-loop systems produced by finite-dimensional compensators designed with each approximation scheme. We have two convergence criteria that indicate how closely a finite-dimensional compensator approximates the infinite-dimensional compensator. One, an obvious criterion, is the frequency response of the compensator, represented by the bode plot. This is a very useful criterion, but it has two disadvantages for computer-aided design applications. First, for a large-order compensator, it can be computationally expensive to compute the frequency response over a wide frequency band; second, while control engineers can glean much useful information from a bode plot, it is not clear what numerical criterion an automated computer-aided design package should use to measure the difference between the frequency responses of two compensators when no engineer is in the design loop.

The alternative convergence criterion for the approximating compensators is the convergence of the control and estimator gains. In finite-dimensional control theory, these gains are matrices, whose dimension varies with the order of the compensator. In the finite-dimensional setting then, there is no obvious way of comparing the gains for compensators of different order. However, in the infinite-dimensional framework of our approximation theory, the gain matrices for each finite-dimensional compensator are transformed into functional control and estimator gains that approximate the corresponding gains for the infinite-dimensional compensator. These functional gains provide a criterion for comparing compensators of different order and compensators designed with different approximation schemes and for measuring the convergence of the finite-dimensional compensators as the order increases. The advantage of the functional gains as a convergence criterion is that they are easy to compute as linear combinations of the basis vectors in the approximation schemes.

We do not prove any convergence theorems in this chapter; we only summarize the pertinent results from [6], which is the theoretical basis for this work. In §3.2, we set up the abstract distributed model—a flexible structure that is the basis for the infinite-dimensional control theory and the approximation theory. Also, we describe the particular hub-beam-tip mass structure to be used here. In §3.3, we define the infinite-dimensional LQG problem for the distributed model of the structure, we discuss the infinite-dimensional compensator that solves the LQG problem, and we discuss the optimal closed-loop system. The infinite-dimensional Riccati operator equations involved in the solution to the infinite-dimensional LQG problem are in §3.3.

Section 3.4 summarizes the approximation theory in [6]. We approximate the structure with a sequence of Ritz–Galerkin approximations in §3.4.1 and then in §§3.4.2 and 3.4.3 define an LQG control problem for each finite-dimensional approximation to the structure. We give the Riccati matrix equations to be solved in each approximating problem and the formulas for computing the functional control and estimator gains after the Riccati equations have been solved. We describe how the solutions to the sequence of finite-dimensional problems converge to the solution to the infinite-dimensional LQG problem. Section 3.4.4 describes the finite-dimensional compensator that results from each finite-dimensional problem and the closed-loop system that consists of the distributed model of the structure and the finite-dimensional compensator.

In §3.5.1, we define the state and control weightings to be used for the optimal linear-quadratic regulator problem for the hub-beam-tip mass structure and discuss the form of the functional control gains for this example. In §3.5.2, we describe the three schemes for approximating the solution to the optimal regulator problem, and we present the numerical results in §3.5.3. This comparison of the three schemes shows how the functional control gains obtained with each approximation scheme converge.

We set up the state estimation problem in §3.6.1 and approximate its solution using each of the approximation schemes. We show the approximating functional estimator gains to demonstrate convergence of the approximating estimators. In §3.6.2, we combine the finite-dimensional control law obtained in §3.5 with the finite-dimensional state estimators to produce the finite-dimensional compensators. We show the bode plots of these compensators to demonstrate convergence to the optimal infinite-dimensional compensator.

For the final comparison of the approximation schemes, in §3.7 we select finite-dimensional compensators that the results in §§3.5 and 3.6 indicate are very close to the infinite-dimensional compensator and form the closed-loop system with a large modal model of the structure. This simulation model includes more than all of the modes that couple with the compensator in the optimal closed-loop system, i.e., the closed-loop system consisting of the distributed model of the structure and the infinite-dimensional compensator. To compare in a third way the structure of the finite-dimensional compensators designed using different approximation schemes and to compare the corresponding closed-loop systems, we examine the closed-loop eigenvalues.

3.2. The Control System.

3.2.1. An Abstract Model of a Flexible Structure. The linear motion of many flexible structures, including the example in this paper, can be represented by the second order (in time) evolution equation

$$\ddot{x} + D_0\dot{x} + A_0x = B_0u, \tag{3.1}$$

where the generalized displacement $x(t)$ is in a real Hilbert space H and the control $u(t)$ is in R^m for some finite integer m. The stiffness operator A_0 is a densely defined nonnegative, selfadjoint linear operator with compact resolvent; the linear damping operator D_0 is nonnegative, symmetric, and bounded relative to A_0; the actuator influence operator B_0 is linear and bounded from R^m to H. The hypotheses on A_0 imply that A_0 has a countably infinite number of eigenvalues, which are real and nonnegative. Furthermore, this sequence of eigenvalues increases without bound, each eigenvalue has finite multiplicity, and the eigenvectors of A_0 are mutually orthogonal and complete in H. (See spectral theory for compact selfadjoint operators [1], [11].) The *natural mode shapes* of the system in (3.1) are the eigenvectors of A_0.

A nontrivial null space of A_0 represents rigid-body motion. In our example and most applications, the rigid-body modes are controllable and the operator

$$\widetilde{A}_0 = A_0 + B_0 B_0^* \tag{3.2}$$

is coercive; i.e., there exists $\rho > 0$ such that

$$\left\langle \widetilde{A}_0 x, x \right\rangle_H \geq \rho |x|_H^2, \qquad x \in D(A_0). \tag{3.3}$$

It is well known that the first-order form of such systems is most naturally studied on an energy space whose norm involves the elastic strain energy associated with the generalized displacement and the kinetic energy associated with the generalized velocity. Hence we define the *strain-energy space* V to be the completion of $D(A_0)$ with respect to the inner product

$$\left\langle \widetilde{A}_0 v_1, v_2 \right\rangle_H = \left\langle \widetilde{A}_0^{1/2} v_1, \widetilde{A}_0^{1/2} v_2 \right\rangle_H. \tag{3.4}$$

Equivalently, V can be defined as $D(\widetilde{A}_0^{1/2})$. The *total-energy space* is $E = V \times H$. Thus $|(x(t), \dot{x}(t))|_E^2$ is twice the total energy in the structure plus $\langle B_0^* x(t), B_0^* x(t) \rangle_{R^m}$. This last term is just the square of the rigid-body rotation in our example and in many applications. The eigenvectors of A_0 are complete in V, as well as in H.

It is shown in [5] and [6] that the operator

$$\mathring{A} = \begin{bmatrix} 0 & I \\ -A_0 & -D_0 \end{bmatrix}, \qquad D(\mathring{A}) = D(A_0) \times D(A_0), \tag{3.5}$$

extends uniquely to an operator A that generates a C_0-semigroup on E. With

$$B = \begin{bmatrix} 0 \\ B_0 \end{bmatrix} \in L(R^m, E) \tag{3.6}$$

and

$$z = \begin{pmatrix} x \\ \dot{x} \end{pmatrix} \in E, \tag{3.7}$$

the first-order form of (3.1) is then

$$\dot{z} = Az + Bu. \tag{3.8}$$

Also, the measurement will have the form

$$y(t) = Cz(t) = [C_1 \quad C_2] \begin{pmatrix} x(t) \\ \dot{x}(t) \end{pmatrix} \tag{3.9}$$

where C is bounded from E to R^p for some finite integer p; i.e., C_1 is bounded from V to R^p, and C_2 is bounded from H to R^p.

TABLE 3.1

Structural data.

r = hub radius	10 in
ℓ = beam length	100 in
I_0 = hub moment of inertial about axis perpendicular to page through 0	100 slug in^2
m_b = beam mass per unit length	.01 slug/in
m_1 = tip mass	1 slug
EI = product of elastic modulus and second moment of cross section for beam	13,333 slug in^3/sec^2
fundamental frequency of undamped structure	.9672 rad/sec

3.2.2. Example. One end of the Euler–Bernoulli beam in Fig. 3.1 is attached rigidly (cantilevered) to a rigid disc which is free to rotate about its center, point 0, which is fixed. Also, a point mass m_1 is attached to the other end of the beam. The control is a torque u applied to the disc, and all motion is in the plane.

The angle θ represents the rotation of the disc (the rigid-body mode), $w(t,s)$ is the elastic deflection of the beam from the rigid-body position, and $w_1(t)$ is the displacement of m_1 from the rigid-body position. The control problem is to stabilize rigid-body motions and linear (small) transverse elastic vibrations about the state $\theta = 0$ and $w = 0$. Our linear model assumes not only that the elastic deflection of the beam is linear but also that the axial inertial force produced by the rigid-body angular velocity has negligible effect on the bending stiffness of the beam. The rigid-body angle need not be small.

For this example, it is a straightforward exercise to derive the three coupled differential equations of motion in θ, w, and w_1. However, as in most applications, we do not need the explicit differential equations; for both theory and computation, we need only the kinetic and strain-energy functionals, the damping functional, and the actuator influence operator.

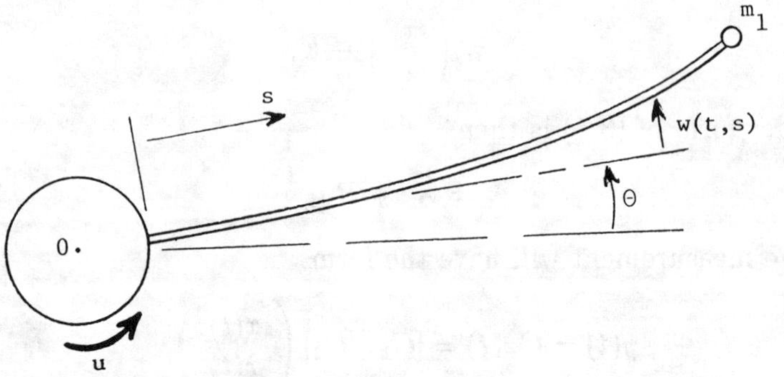

FIG. 3.1. *Flexible structure.*

The generalized displacement vector is

(3.10) $$x = (\theta, w, w_1) \in H_0 = R \times L_2(0, \ell) \times R.$$

The space H_0 has the ordinary inner product for a product space, and it is useful to define a second inner product on H_0. The kinetic energy in the system is

(3.11) $$\text{Kinetic Energy } = \tfrac{1}{2}\langle \dot{x}, \dot{x}\rangle_H$$

where H is H_0 with the inner product

(3.12) $$\begin{aligned}\langle x, \widehat{x}\rangle_H = m_b \int_0^\ell [w(s) + (r+s)\theta]\,[\widehat{w}(s) + (r+s)\widehat{\theta}]\,ds \\ +I_0\theta\widehat{\theta} + m_1[w_1 + r + \ell)\theta]\,[\widehat{w}_1 + (r+\ell)\widehat{\theta}],\end{aligned}$$

for $x = (\theta, w, w_1)$ and $\widehat{x} = (\widehat{\theta}, \widehat{w}, \widehat{w}_1) \in H_0$. We need not write it explicitly, but there exists a unique selfadjoint linear mass operator M_0 on H_0 such that

(3.13) $$\langle x, \widehat{x}\rangle_H = \langle M_0 x, \widehat{x}\rangle\, H_0.$$

It is easy to see that M_0 is bounded and coercive. Hence, H_0 and H have equivalent norms.

The only strain energy is in the beam and is given by

(3.14) $$\text{Strain Energy } = \tfrac{1}{2}a(x, x)$$

with

(3.15) $$a(x, \widehat{x}) = EI \int_0^\ell w''\widehat{w}''\,ds.$$

where $(\cdot)'' = \partial^2(\cdot)/\partial s^2$. To make $a(\cdot,\cdot)$ into an inner product, we must have

(3.15) $$a(x, \widehat{x}) = EI \int_0^\ell w''\widehat{w}''\,ds.$$

where $(\cdot)'' = \partial^2(\cdot)/\partial s^2$. To make $a(\cdot,\cdot)$ into an inner product, we must account for rigid-body rotation. Thus we set

(3.16) $$\langle x, \widehat{x}\rangle_V = a(x, \widehat{x}) + \theta\widehat{\theta}$$

and define

(3.17) $$V = \left\{x = (\theta, \phi, \phi(\ell)) : \phi \in H^2(0, \ell), \phi(0) = \phi'(0) = 0\right\}.$$

The strain-energy functional $a(\cdot,\cdot)$ determines the stiffness operator A_0 for (3.1); i.e., there exists a unique nonnegative selfadjoint linear operator A_0 with dense domain for which

(3.18) $$a(x, \widehat{x}) = \langle A_0 x, \widehat{x}\rangle_H, | x, \widehat{x} \in D(A_0).$$

With the imbedding $V \subset H = H' \subset V', D(A_0)$ is the image of H under the inverse Riesz map of V. See [6, §2], [17, Chap. VI].

To write the equations of motion in the form (3.1) with A_0 selfadjoint, we must use the space H. In H_0, the mass operator M_0 sits in front of x when A_0 is H_0-selfadjoint and D_0 is H_0-symmetric. In other words, the difference between the operators A_0, D_0, and B_0 for H_0 and the corresponding operators for H is that the latter are the former followed by M_0^{-1}. It would be inconvenient for our approximation theory to have to carry the mass operator around in the notation, and in applications it is inconvenient to write down M_0 explicitly. Hence we let the H-inner product take care of the mass distribution, so that all we need to write explicitly is the kinetic energy.

Were we to write the equations of motion for this example with M_0 in front of $\ddot{x}$, the input operator B_0 would be

$$B_0 = (1,0,0). \tag{3.19}$$

The B_0 for (3.1)–(3.6) is then $(M_0^{-1}B_0)$, and it is important to note that, if $(M_0^{-1}B_0)^{*H}$ is the H-adjoint of $(M_0^{-1}B_0)$ and B_0^* is the H_0-adjoint of B_0, then

$$(M_0^{-1}B_0)^{*H} = B_0^*, \tag{3.20}$$

so that

$$\begin{aligned} \langle x, \hat{x} \rangle_V &= a(x, \hat{x}) + \langle B_0 B_0^* x, \hat{x} \rangle_{H_0} \\ &= a(x, \hat{x}) + \langle (M_0^{-1}B_0)(M_0^{-1}B_0)^{*H} x, \hat{x} \rangle >_H. \end{aligned} \tag{3.21}$$

We assume that the beam has Voigt–Kelvin viscoelastic damping, so that the damping operator in (3.1) is

$$D_0 = c_0 A_0 \tag{3.22}$$

where $c_0 = 10^{-4}$.

3.3. The Infinite-Dimensional LQG Problem.

3.3.1. The Optimal Control Problem. The deterministic optimal linear regulator problem is to choose $u \in L_2(0, \infty; R^m)$ to minimize

$$J(z(0), u) = \int_0^\infty (\langle Qz(t), z(t) \rangle_E + \langle Ru(t), u(t) \rangle_{R^m})\, dt. \tag{3.23}$$

where $z(t)$ is the state vector in (3.8), Q is a nonnegative selfadjoint bounded linear operator on E, and R is a positive definite symmetric $m \times m$ matrix. See [5], [6] for more detail on this problem for flexible structures. According to §3 of [6], a sufficient condition for a solution to this problem to exist is that $(D_0 + B_0 B_0^*)$ be coercive, as it is in the example in this chapter. That the optimal control is unique follows from the fact that, since R is positive definite, $J(z(0), u)$ is a coercive function of u.

The optimal control has the feedback form

$$u(t) = -Fz(t) \tag{3.24}$$

where

$$F = R^{-1}B^*\Pi \tag{3.25}$$

and Π is a nonnegative selfadjoint bounded linear operator on E that satisfies the Riccati equation

$$A^*\Pi + \Pi A - \Pi B R^{-1} B^* \Pi + Q = 0. \tag{3.26}$$

A sufficient condition for existence of a bounded nonnegative selfadjoint solution to (3.26) is that $(A_0 + B_0B_0^*)$ and $(D_0 + B_0B_0^*)$ be coercive, as both are in the example here. A sufficient condition for a nonnegative selfadjoint bounded solution to (3.26) to be unique is that, for each initial condition $z(0)$, any control that makes the performance index finite drives $z(t)$ to zero asymptotically [6]. For flexible structures this usually means that any unstable modes of the system (Q, A) are observable; i.e., any unstable modes of the open-loop structure are observable with the output Qz. Coercivity of $(A_0 + B_0B_0^*)$ and $(D_0 + B_0B_0^*)$ and observability of unstable modes of (Q, A) guarantee that the closed-loop semigroup, generated by $(A - BF)$, is uniformly exponentially stable.

When $m = 1$ (i.e., one actuator), the operators B_0 and B are elements of H and E, respectively, and the control law in (3.24) has the form

$$u(t) = -\langle f, x(t)\rangle_V - \langle g, \dot{x}(t)\rangle_H \tag{3.27}$$

with $f \in V$ and $g \in H$ given by

$$(f, g) = R^{-1}\Pi B. \tag{3.28}$$

The generalization to m actuators is trivial (see [6]), but the notation is more complicated. We call f and g *functional control gains.*

3.3.2. The Estimator. For the control system in (3.8) and (3.9), we define the state estimator

$$\dot{\hat{z}} = A\hat{z} + Bu + F(y - C\hat{z}) \tag{3.29}$$

with $\hat{F} \in L(R^p, E)$. For the estimator error

$$e(t) = z(t) - \hat{z}(t), \tag{3.30}$$

we have

$$e(t) = \hat{S}(t)e(0), \qquad t \geq 0, \tag{3.31}$$

where $\hat{S}(t)$ is the semigroup generated by $(A - \hat{F}C)$. Thus, if $\hat{S}(t)$ is uniformly exponentially stable, the estimator error decays exponentially.

We call the estimator in (3.27) optimal when

(3.32) $$\widehat{F} = \widehat{\Pi} C^* \widehat{R}^{-1}$$

with $\widehat{\Pi} \in L(E, E)$ the nonnegative selfadjoint solution to the Riccati equation

(3.33) $$A\widehat{\Pi} + \widehat{\Pi} A^* - \widehat{\Pi} C^* \widehat{R}^{-1} C \widehat{\Pi} + \widehat{Q} = 0$$

for some nonnegative selfadjoint $\widehat{Q} \in L(E, E)$ and positive definite symmetric $p \times p$ matrix $\widehat{R}$. The justification for calling this estimator optimal is that it is the optimal filter for the distributed model of the structure if a stationary Gaussian white noise process with covariance operator $\widehat{Q}$ is added to the right side of (3.8) and a stationary Gaussian white noise process with covariance matrix $\widehat{R}$ is added to the right side of (3.9). (See [1], [4] for infinite-dimensional filtering.) However, in many engineering applications, $\widehat{Q}$ and $\widehat{R}$ are chosen to produce desired characteristics of the transient response of the estimator, such as stability margin and robustness, instead of being chosen to represent modeled noise processes.

For $p = 1$ (i.e., one measurement), there exist vectors $c_1 \in V$ and $c_2 \in H$ such that

(3.34) $$Cz = \langle c_1, x \rangle_V + \langle c_2, \dot{x} \rangle_H .$$

Also, the estimator gain in (3.32) can be written

(3.35) $$\widehat{F} = (\widehat{f}, \widehat{g})$$

where $\widehat{f} \in V$ and $\widehat{g} \in H$ are *functional estimator gains*, which are given by

(3.36) $$(\widehat{f}, \widehat{g}) = \widehat{R}^{-1} \widehat{\Pi}(c_1, c_2).$$

3.3.3. The Optimal Compensator and Closed-Loop System. The optimal compensator consists of the optimal estimator and the control law in (3.24) applied to the output of this estimator, rather than to the state vector (which is not available for feedback). Thus the control applied to the structure is

(3.37) $$u(t) = -F\widehat{z}(t).$$

The optimal compensator has the transfer function

(3.38) $$\Phi(s) = F(sI - [A - BF - \widehat{F}C])^{-1} \widehat{F},$$

which is an $m \times p$ matrix function of the complex variable s. Since E has infinite dimension, this transfer function is irrational in general.

The resulting closed-loop system is

(3.39) $$\begin{pmatrix} z(t) \\ \widehat{z}(t) \end{pmatrix} = S_{\infty,\infty}(t) \begin{pmatrix} z(0) \\ \widehat{z}(0) \end{pmatrix}, \qquad t \geq 0,$$

where $S_{\infty,\infty}(t)$ is the semigroup generated on $E \times E$ by the operator

$$(3.40)\quad A_{\infty,\infty} = \begin{bmatrix} A & -BF \\ \widehat{F}C & [A - BF - \widehat{F}C] \end{bmatrix}, \qquad D(A_{\infty,\infty}) = D(A) \times D(A).$$

It is easy to show that

$$(3.41)\qquad \sigma(A_{\infty,\infty}) = \sigma(A - BF) \cup \sigma(A - \widehat{F}C),$$

where $\sigma(A_{\infty,\infty})$ is the spectrum of $(A_{\infty,\infty})$. (See Fig. 3.2.)

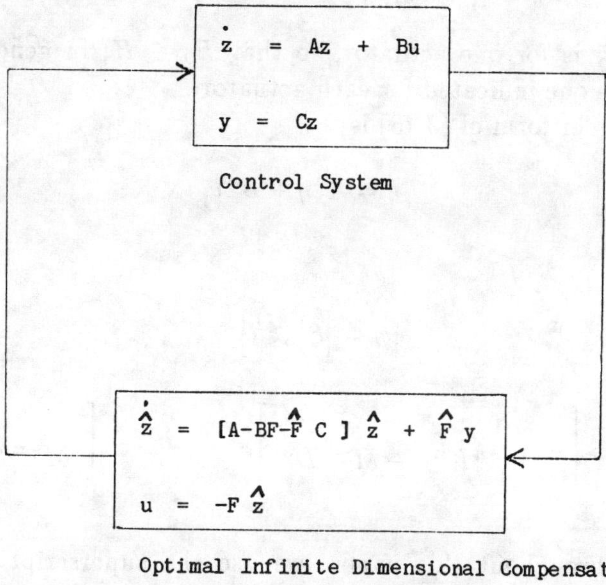

FIG. 3.2. *Optimal closed-loop system.*

3.4. Approximation.

3.4.1. Approximation of the Structure. For the rest of the paper, we assume the following hypothesis.

Hypothesis 3.4.1. There exists a sequence of finite-dimensional subspaces V_n of V such that the sequence of orthogonal projections P_{Vn} converge strongly in V to the identity, where P_{Vn} is the V-projection onto V_n. Also, each V_n is the span of n linearly independent vectors e_j.

The basis vectors may change from one V_n to the next, as in most finite element schemes. We will approximate the total energy space $E = V \times H$ by $E_n = V_n \times V_n$, which has the same inner product as E. We denote the E-projection onto E_n by P_{En}.

The Ritz–Galerkin approximation of the control system is based on approximating the generalized displacement of the structure by

$$(3.42)\qquad x_n(t) = \sum_{j=1}^{n} \xi_j(t) e_j,$$

where $\xi(t) = (\xi_1(t), \xi_2(t), \ldots, \xi_n(t))^T$ satisfies

$$M^n\ddot{\xi}(t) + D^n\dot{\xi}(t) + K^n\xi(t) = B_0^n u(t), \tag{3.43}$$

and the mass matrix M^n, damping matrix D^n, stiffness matrix K^n, and actuator influence matrix B_0^n are given by

$$\begin{aligned} M^n &= [\langle e_i, e_j\rangle_H], \qquad D^n = [\langle D_0 e_i, e_j\rangle_H], \\ K^n &= [\langle e_i, e_j\rangle_V] - [\langle B_0^* e_i, B_0^* e_j\rangle_{R^m}], \\ B_0^n &= [\langle e_i, B_0\rangle_H]. \end{aligned} \tag{3.44}$$

The matrix B_0^n is for one actuator, so that $B_0 \in H$; in general, B_0^n has a column like the one indicated for each actuator.

The first-order form of (3.43) is

$$\dot{\eta} = A^n\eta + B^n u \tag{3.43$'$}$$

where

$$\eta = \left[\xi^T, \dot{\xi}^T\right]^T \tag{3.45}$$

and

$$A^n = \begin{bmatrix} 0 & I \\ -M^{-n}K^n & -M^{-n}D^n \end{bmatrix}, \qquad B^n = \begin{bmatrix} 0 \\ M^{-n}B_0^n \end{bmatrix} \tag{3.46}$$

Note. Throughout this paper, we use the superscript n in the designation of matrices in the nth approximating system and control problem, like A^n, B^n, M^n, etc. Hence, the superscript n indicates the order of approximation—and it *never* indicates a power of the matrix. By M^{-n}, we denote the inverse of the mass matrix M^n.

For the approximating LQG problems, we need the following matrices:

$$\widetilde{K}^n = [\langle e_i, e_j\rangle_V] = K^n + [\langle B_0^* e_i, B_0^* e_j\rangle_{R^m}]. \tag{3.47}$$

and

$$W^n = \begin{bmatrix} \widetilde{K}^n & 0 \\ 0 & M^n \end{bmatrix}. \tag{3.48}$$

3.4.2. The Approximating Optimal Control Problems. The nth optimal control problem is: given $\eta(0) \in R^{2n}$, choose $u \in L_2(0, \infty; R^m)$ to minimize

$$J_n(\eta(0), u) = \int_0^\infty (\eta(t)^T \widetilde{Q}^n \eta(t) + u(t)^T R u(t))\, dt \tag{3.49}$$

where $\eta(t)$ is the vector in (3.43$'$) and (3.45) and $\widetilde{Q}^n$ is a symmetric nonnegative $2n \times 2n$ matrix defined as follows. Since Q is a selfadjoint bounded linear operator on $E = V \times H$, we can partition Q as

$$(3.50) \qquad Q = \begin{bmatrix} Q_0 & Q_1^* \\ Q_1^* & Q_2^* \end{bmatrix}$$

where $Q_0 \in L(V), Q_1 \in L(H,V)$, and $Q_2 \in L(H)$, with Q_0 and Q_2 selfadjoint and nonnegative. Similarly, we partition $\widetilde{Q}^n$ into $n \times n$ blocks as

$$(3.51) \qquad \widetilde{Q}^n = \begin{bmatrix} \widetilde{Q}_0^n & \widetilde{Q}_1^n \\ \widetilde{Q}_1^{nT} & \widetilde{Q}_2^{n2} \end{bmatrix}$$

with

$$(3.52) \qquad \widetilde{Q}_1^n = [\langle e_i, Q_0 e_j \rangle_V]\,, \quad \widetilde{Q}_1^n = [\langle e_i, Q_1 e_j \rangle_V]\,, \quad \widetilde{Q}_2^n = [\langle e_i, Q_2 e_j \rangle_V]\,.$$

Note that, if $\langle Qz, z\rangle_E$ is twice the total energy (kinetic plus potential) in the structure plus the square of any rigid-body displacement, then $\widetilde{Q}^n$ is the matrix W^n in (3.48).

The matrix $\widetilde{Q}^n$ is related to the nth approximation to the operator Q by

$$(3.53) \qquad \widetilde{Q}^n = W^n Q^n$$

where $\widetilde{Q}^n$ is the matrix representation of $(P_{En} Q)|_{En}$. Since Q is selfadjoint, $\widetilde{Q}^n$ is symmetric, but Q^n in general is not.

The optimal control law for the nth optimal control problem is

$$(3.54) \qquad u_n = -F^n \eta(t)$$

where

$$(3.55) \qquad F^n = R^{-1} B^{nT} \widetilde{\Pi}^n$$

and $\widetilde{\Pi}^n$ is the minimal nonnegative, symmetric solution to the Riccati matrix equation

$$(3.56) \qquad (A^n)^T \widetilde{\Pi}^n + \widetilde{\Pi}^n A^n - \widetilde{\Pi}^n B^n R^{-1} (B^n)^T \widetilde{\Pi}^n + \widetilde{Q}^n = 0.$$

A sufficient condition for such a solution to exist is that the system in (3.43′) be controllable. Another sufficient condition is that $(D_0 + B_0 B_0^*)$ be coercive.

The approximation theory for the optimal control problem is based on the approximation of the solution to the infinite-dimensional Riccati equation (3.26) via the solution of the finite-dimensional Riccati equation (3.56). If we define Π_n to be the operator on E^n whose matrix representation is

$$(3.57) \qquad \Pi^n = W^{-n} \widetilde{\Pi}^n$$

(recall that W^{-n} is the inverse of W_n), then $\Pi_n P_{En}$ converges to Π as discussed in [6, §5]. Here, let us just say that if $(D_0 + B_0 B_0^*)$ is H-coercive and Q is E-coercive, then $\Pi_n P_{En}$ converges strongly to Π.

To study the convergence of the control law in (3.54) and (3.55) to the control law in (3.24) and (3.25), we will translate the nth control law into an approximation to the functional control gains in (3.27) and (3.28). We continue to treat only the single-actuator case here for simplicity. In view of (3.42) and (3.45), (3.54) is equivalent to

$$u_n(t) = -\langle f_n, x_n(t)\rangle_V - \langle g_n, \dot{x}_n(t)\rangle_H \tag{3.58}$$

with the nth functional control gains $f_n, g_n \in V_n$. If we write

$$f_n = \sum_{j=1}^{n} \beta_j^f e_j \quad \text{and} \quad g_n = \sum_{j=1}^{n} \beta_j^g e_j \tag{3.59}$$

with $\beta^f, \beta^g \in R^n$, and then equate (3.54) and (3.58) using (3.42) and (3.45), we obtain

$$\begin{pmatrix} \beta^f \\ \beta^g \end{pmatrix} = W^{-n}(F^n)^T \tag{3.60}$$

where F^n is given by (3.55).

Thus, to compute the nth approximation to the optimal control law, we solve the Riccati matrix equation (3.56), compute the feedback gain matrix according to (3.55), and compute the approximating functional gains according to (3.59) and (3.60). If is shown in [6] that, if $\Pi_n P_{En}$ converges strongly to Π, then f_n and g_n converge in V and H, respectively, to the f and g in (3.27).

3.4.3. The Approximating Estimators. We denote by C^n the matrix representation of $C|_{En}$, where C is the measurement operator in (3.9). Also, we denote by $\widehat{Q}^n$ the matrix representation of $\widehat{Q}|_{En}$, where $\widehat{Q}$ is the operator in (3.33), and we define

$$\tilde{\widehat{Q}}^n = \widehat{Q}^n W^{-n}. \tag{3.61}$$

Since $\widehat{Q}$ is selfadjoint and nonnegative, $\tilde{\widehat{Q}}^n$ is symmetric and nonnegative.

The nth state estimator is

$$\dot{\widehat{\eta}} = A^n\widehat{\eta} + B^n u + \widehat{F}^n(y - C^n\widehat{\eta}) \tag{3.62}$$

where the $2-n$ vector $\widehat{\eta}(t)$ is an estimate of the $\eta(t)$ in (3.43′). The gain matrix $\widehat{F}^n$ is

$$\widehat{F}^n = \tilde{\widehat{\Pi}}^n (c^n)^T \widehat{R}^{-1} \tag{3.63}$$

with $\tilde{\widehat{\Pi}}^n$ the nonnegative symmetric solution to the Riccati matrix equation

$$A^n\tilde{\widehat{\Pi}}^n + \tilde{\widehat{\Pi}}^n(A^n)^T - \tilde{\widehat{\Pi}}^n(C^n)^T\widehat{R}^{-1}C^n\tilde{\widehat{\Pi}}^n + \tilde{\widehat{Q}}^n = 0. \tag{3.64}$$

The approximation theory for the estimator problem is similar to that for the control problem. See [6, §8.4]. If $\widehat{\Pi}_n$ is the operator on E_n whose matrix representation is

$$\widehat{\Pi}^n = \tilde{\widehat{\Pi}}^n W^n, \tag{3.65}$$

then $\widehat{\Pi}_n P_{En}$ converges to $\widehat{\Pi}$, the solution to (3.33), under suitable conditions. (Yes, (3.65) is different from (3.57).) For example, it follows from [6, Thm. 8.5] that $\widehat{\Pi}_N P_{En}$ converges strongly to $\widehat{\Pi}$ if A_0 and D_0 are coercive and the system $(A, \widehat{Q})$ is controllable. When rigid-body modes are present (i.e., A_0 is not coercive), the proven convergence is more complicated but essentially similar to that for the control problem.

In particular, the estimator gain matrix $\widehat{F}^n$ is the matrix representation of an operator that approximates the gain operator $\widehat{F}$ in (3.29) and (3.32). For a single measurement ($p = 1$), we can represent $\widehat{F}^n$ with a pair of functional estimator gains $\widehat{f}_n$ and $\widehat{g}_n$ that approximate those in (3.35). If we write

$$\widehat{F}^n = \begin{pmatrix} \beta^f \\ \beta^g \end{pmatrix}, \tag{3.66}$$

then

$$\widehat{f}_n = \sum_{i=1}^{n} \beta_j^f e_j \quad \text{and} \quad \widehat{g}_n = \sum_{i=1}^{n} \beta_j^g e_j. \tag{3.67}$$

When $\widehat{\Pi}_n P_{En}$ converges to $\widehat{\Pi}$ strongly, $\widehat{f}_n$ and $\widehat{g}_n$ converge in V and H, respectively, to $\widehat{f}$ and $\widehat{g}$.

3.4.4. The Finite-Dimensional Compensators and Realizable Closed-Loop Systems. The nth compensator consists of the estimator in (3.62) and the control law

$$u_n = -F^n \widehat{\eta} \tag{3.68}$$

where F^n is the control gain matrix given by (3.55). The nth compensator has the transfer function

$$\Phi_n(s) = F^n \left(sI - \left[A^n - B^n F^n - \widehat{F}^n C^n \right] \right)^{-1} \widehat{F}^n, \tag{3.69}$$

which is an $m \times p$ matrix function of the complex variable s for each n.

When the nth compensator is applied to the structure in §§3.2 and 3.3, the closed-loop system in Fig. 3.3 results.

The nth closed-loop system satisfies the differential equation

$$\begin{pmatrix} \dot{z} \\ \dot{\widehat{\eta}} \end{pmatrix} = A_{\infty,n} \begin{pmatrix} z \\ \widehat{\eta} \end{pmatrix} \tag{3.70}$$

where the operator

$$A_{\infty,n} = \begin{bmatrix} A & -BF^n \\ \widehat{F}^n C & [A^n - B^n F^n - \widehat{F}^n C^n] \end{bmatrix}, \quad D(A_{\infty,n}) = D(A) \times R^{2n}, \tag{3.71}$$

generates the closed-loop semigroup $S_{\infty,n}(t)$ on $E \times R^{2n}$.

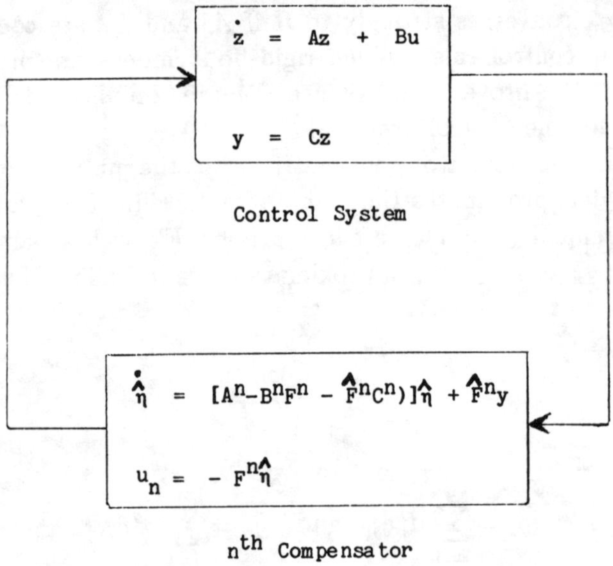

FIG. 3.3. *nth Closed-loop system.*

Section 9 of [6] contains theorems describing how $A_{\infty,n}$ and $S_{\infty,n}(t)$ converge, respectively, to the $A_{\infty,\infty}$ and $S_{\infty,\infty}(t)$ in (3.41) and (3.40) as n increases, and how the transfer function of the finite-dimensional compensator converges to the transfer function in (3.39) of the optimal infinite-dimensional compensator. An easy result says that, when the nth functional control and estimator gains converge to the functional control and estimator gains for the infinite-dimensional compensator, then $S_{\infty,n}(t)$ converges strongly to $S_{\infty,\infty}(t)$ and the convergence is uniform in t for t in bounded intervals. This is not as useful as it may seem though; it does not guarantee that, if the optimal closed-loop system (represented by $S_{\infty,\infty}(t)$) is uniformly exponentially stable, then so is the nth closed-loop system for n sufficiently large. We have proved this stability result only when the basis vectors e_j are the natural mode shapes of free vibration and the structural damping does not couple the modes, although our numerical evidence suggests that the result is true under more general conditions.

3.5. The Optimal Control Problem and the Approximations for the Example.

3.5.1. The Optimal Control Problem. We take $Q = I$ in the performance index in (3.23). This means that the state weighting term $\langle Qz, z\rangle_E$ is twice the total energy in the structure plus the square of the rigid-body rotation. Since there is one input, the control weighting R is a scalar.

According to (3.27), the optimal control has the feedback form

$$u(t) = -\langle f, x(t)\rangle_V - \langle g, \dot{x}(t)\rangle_H \tag{3.72}$$

where $x(t)$ has the form (3.10) and

$$f = (\alpha_f, \phi_f, \beta_f) \in V, \qquad g = (\alpha_g, \phi_g, \beta_g) \in H. \tag{3.73}$$

Note that $\beta_f = \phi_f(\ell)$ is not used in the control law—recall (3.15) and (3.16).

3.5.2 The Three Approximation Schemes.

3.5.2.1. Cubic Hermite splines. Our first approximation of the distributed model of the structure is based on a finite element approximation of the beam that uses cubic Hermite splines as basis functions [16], [18]. These splines and their first derivatives are continuous at the nodes, which are evenly spaced for the results presented here. Because the basis vectors e_j in the approximation scheme in §3.4 must be in the space V defined in (3.17), we write them as

$$e_1 = (1, 0, 0), \tag{3.74a}$$

$$e_j = (0, \phi_j, \phi_j(1)), \qquad j = 2, 3, \cdots, \tag{3.74b}$$

where the ϕ_j's are the cubic splines and n is the number of splines. For n_e elements, there are $2n_e$ linearly independent splines. The $2n_e$ elastic degrees of freedom usually are taken to be the displacements and slopes at the nodes, relative to the rigid-body position. With the rigid-body mode, the total number of degrees of freedom with Hermite splines is $n = 2n_e + 1$.

For the numerical solution to the optimal control problem, we have only to plug into the formulas of §3.4. The matrices in (3.44) are calculated according to (3.12), (3,15), and (3.16), with B_0 in (3.44) equal to M_0^{-1} following the B_0 in (3.19). In particular,

$$K^n = [a(e_i, e_j)], \qquad D^n = c_0 K^n, \tag{3.75}$$

$$\begin{aligned} B_0^n &= [1 \;\; 0 \;\; 0 \;\; \ldots \;\; 0]^T \\ &= \left[\langle e_i, M_0^{-1}(1,0,0)\rangle_H\right] = \langle e_i, (1,0,0)\rangle_{H_0}\right]. \end{aligned} \tag{3.76}$$

Note that the first row and column of K^n are zero. The matrix $\widetilde{K^n}$ in (3.47) is K^n with 1 added to the first element. The matrices A^n and $B^n A$ are given by (3.46) and, since $Q = I$, the matrix $\widetilde{Q}^n$ is the W^n in (3.48).

With these matrices, we solve the Riccati equation (3.56) and compute the control gain matrix according to (3.55). Then, we use (3.59) and (3.60) to compute the approximations to the functional gains, which are

$$f_n = (\alpha_{fn}, \phi_{fn}, \beta_{fn}), \qquad g_n = (\alpha_{gn}, \phi_{gn}, \beta_{gn}). \tag{3.77}$$

For numerical solution of the Riccati matrix equation, we have used the standard eigenvector decomposition of the Hamiltonian matrix [12], usually called the Potter method.

3.5.2.2. Cubic B-splines. The B-splines [16], [18] are similar to the Hermite splines, except that the B-splines have continuous second derivatives, as well as continuous displacements and slopes. We can think of deriving the B-splines from the Hermite splines by imposing the linear constraint requiring continuous second derivative (bending moment) at each node. These n_e constraints leave n_e degrees of freedom for the beam approximation and $n = n_e + 1$ degrees of freedom for the structure, instead of the $2n_e + 1$ with the Hermite splines.

As usual, we take the B-spline degrees of freedom for the beam to be the nodal displacements relative to the rigid-body position. The basis vectors for our approximation scheme again have the form (3.74), and the formulas in (3.75), (3.76) are the same (but we use the new e_i's). Again, after solving (3.55), we use (3.59) and (3.60) to compute the approximating functional gains, which have the form (3.77).

3.5.2.3. Natural modes. In our third scheme for approximating the solution to the infinite-dimensional control problem, we use the natural mode shapes of free undamped vibration as the basis vectors e_1. The first natural mode is the rigid-body mode $e_1 = (1, 0, 0)$ and the others are flexible modes having the form

$$e_j = (\theta_j, \phi_j, \phi_j(\ell)), \qquad j = 2, 3, \cdots, \qquad n = n_m + 1, \tag{3.78}$$

where each θ_j is nonzero and n_m is the number of flexible modes used. To obtain the flexible modes, we used 24 elements with the Hermite splines to approximate the eigenvalue problem for the stiffness operator A_0. This gave convergence for the first twelve natural mode shapes and frequencies, which are more than we need. This scheme for approximating the solution to the control problem is then a two-stage scheme in that the modes must be approximated before they can be used in the approximating control problems.

The stiffness matrix K^n in this scheme is the diagonal matrix with the squares of the natural frequencies on the diagonal, including zero as the first element. We normalized the mode shapes with respect to mass, so that the mass matrix is the identity matrix. The ith term in B_0^n is the amount of hub rotation in the ith mode shape. We still solve the Riccati equation (3.55) and use (3.59) and (3.60) to compute the approximating functional gains. This time, each e_j in (3.59) is a mode shape that must be evaluated as a linear combination of Hermite splines, but this is just bookkeeping.

3.5.3. Numerical Results. For the damping coefficient $c_0 = 10^{-4}$ and the control weighting $R = .05$, Figs. 3.4a and 3.4b show the functional gain kernels ϕ''_{fn} and ϕ_{gn} computed using cubic Hermite splines for $n_e = 4, 6, 8$

and 10 beam elements. Table 3.2 lists the corresponding scalar components of the gains. Figures 3.5a and 3.5b and Table 3.3 show the functional gains computed using cubic B-splines for n_e = 4, 6, 8, 10 and 12 elements, and Figs. 3.6a and 3.6b and Table 3.4 show the functional gains computed using $n_m = 4, 6, 8, 10$ and 12 flexible system mode shapes as basis functions for the beam approximation.

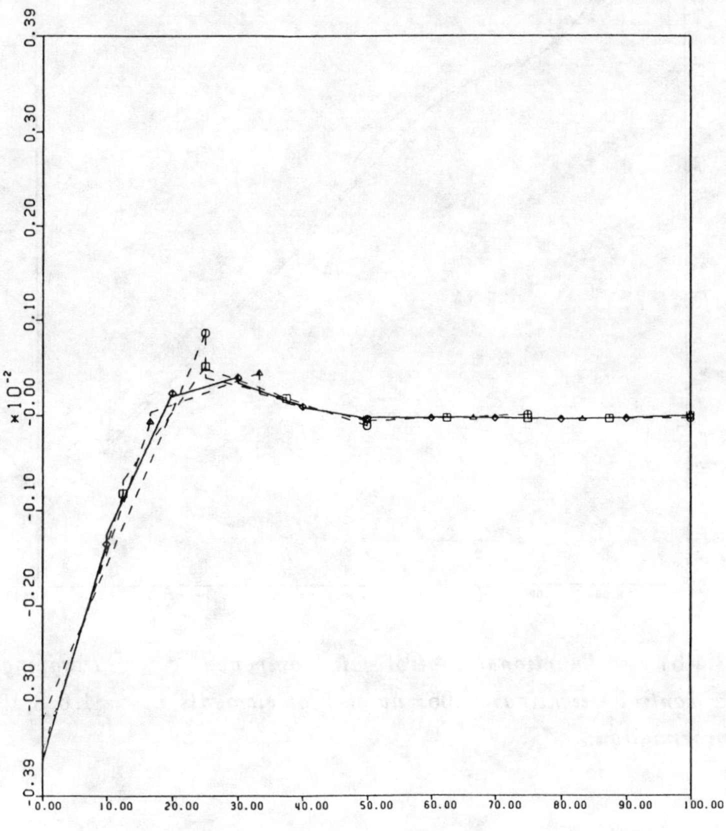

FIG. 3.4(a). *Functional control gain component ϕ''_{fn}. Damping coefficient* $c_0 = 10^{-4}$; *control weighting* $R = .05$; *number of elements* $n_e = 4, 6, 8, 10$. *Hermite Spline Approximation.*

We have plotted ϕ''_{fn} because the second derivative appears in the strain-energy inner product and because ϕ_{fn} converges in $H^2(0, \ell)$, so that ϕ''_{fn} converges in L_2. The numerical results indicate that ϕ''_{fn} and ϕ_{gn} converge uniformly on $[0, \ell]$ for all three of our approximation schemes, even though the Hermite splines yield a ϕ''_{fn} that is discontinuous at the nodes for each n. We have seen this convergence for the functional gains corresponding to beams in all of our structure-control examples, and we suspect that it is the case generally, although the convergence theorems in [6] guarantee only L_2-convergence for ϕ_{gn} and ϕ''_{fn}. Since β_{fn} is not used in the nth control law, we omit it from the tables.

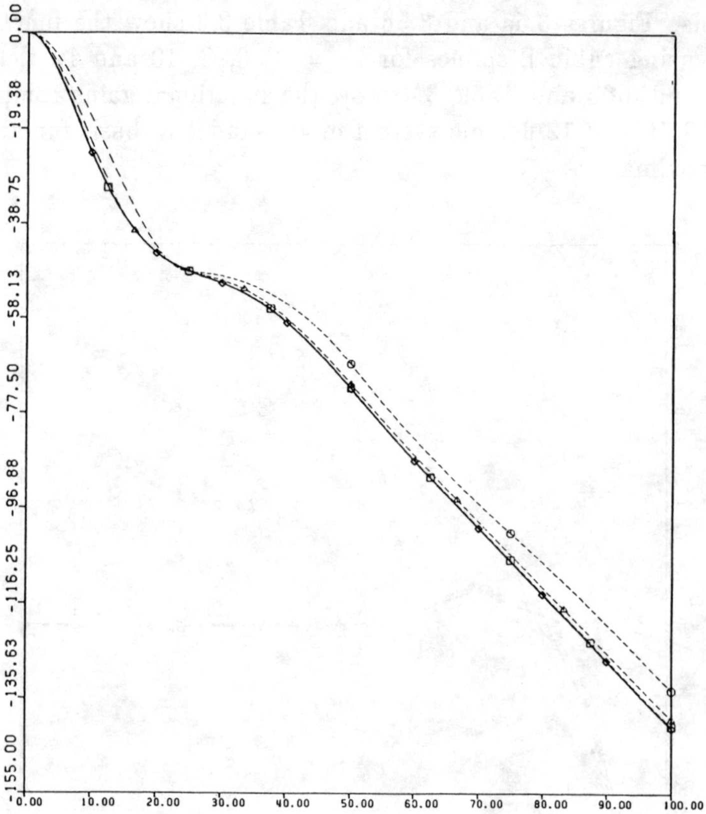

FIG. 3.4(b). *Functional control gain component ϕ_{gn}. Damping coefficient $c_0 = 10^{-4}$; control weighting = .05; number of elements $n_e = 4, 6, 8, 10$. Hermite Spline Approximation.*

TABLE 3.2

Scalar components of functional control gains. Damping coefficient $c_0 = 10^{-4}$; control weighting $R = .05$; number of elements $n_e = 4, 6, 8, 10$. Hermite Spline Approximation.

n_e	α_{fn}	α_{gn}	β_{gn}
4	4.4721	1.2440	−133.87
6	4.4721	1.2973	−139.69
8	4.4721	1.3106	−141.15
10	4.4721	1.3141	−141.54

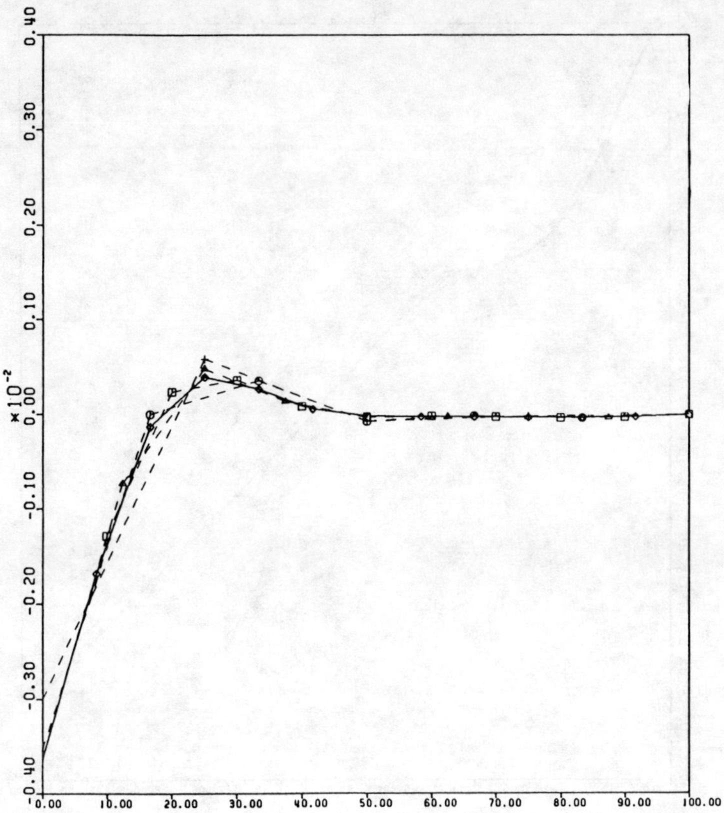

FIG. 3.5(a). *Functional control gain component ϕ''_{fn}. Damping coefficient $c_0 = 10^{-4}$; control weighting $R = .05$; number of elements $n_e = 4, 6, 8, 10, 12$. B-Spline Approximation.*

TABLE 3.3

Scalar components of functional control gains. Damping coefficient $c_0 = 10^{-4}$; control weighting $R = .05$; number of elements $n_e = 4, 6, 8, 10, 12$. B-Spline Approximation.

n_e	α_{fn}	α_{gn}	β_{gn}
4	4.4721	1.1366	−121.87
6	4.4721	1.2817	−137.98
8	4.4721	1.3076	−140.85
10	4.4721	1.3135	−141.46
12	4.4721	1.3151	−141.65

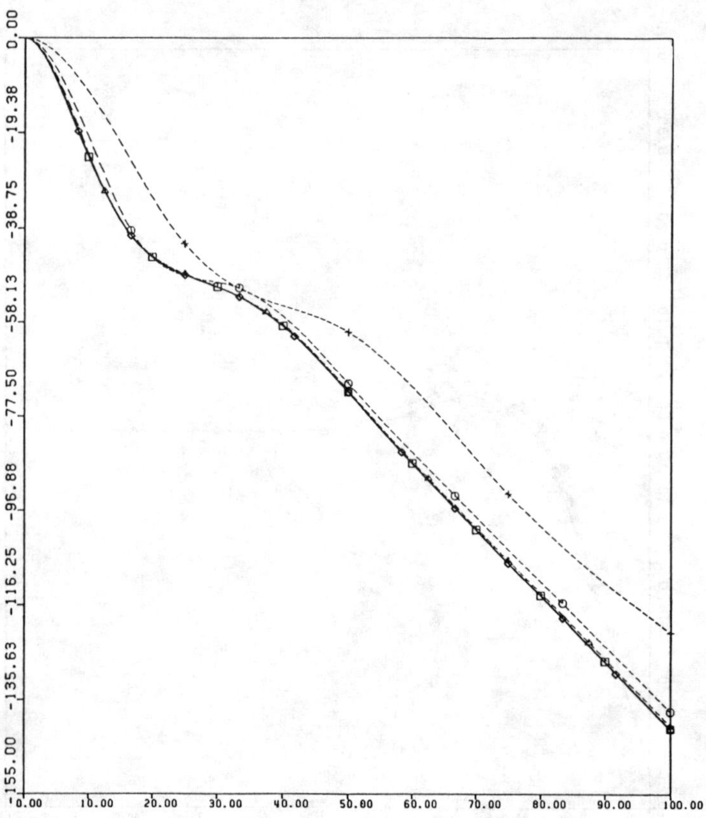

FIG. 3.5(b). *Functional control gain component ϕ_{gn}. Damping coefficient $c_0 = 10^{-4}$; control weighting $R = .05$; number of elements $n_e = 4, 6, 8, 10, 12$. B-Spline Approximation.*

TABLE 3.4

Scalar components of functional control gains. Damping coefficient $c_0 = 10^{-4}$; control weighting $R = .05$; number of elements $n_m = 4, 6, 8, 10, 12$. Modal Approximation.

n_e	α_{fn}	α_{gn}	β_{gn}
4	4.4721	1.1191	−119.78
6	4.4721	1.2817	−137.90
8	4.4721	1.3093	−140.99
10	4.4721	1.3143	−141.56
12	4.4721	1.3157	−141.71

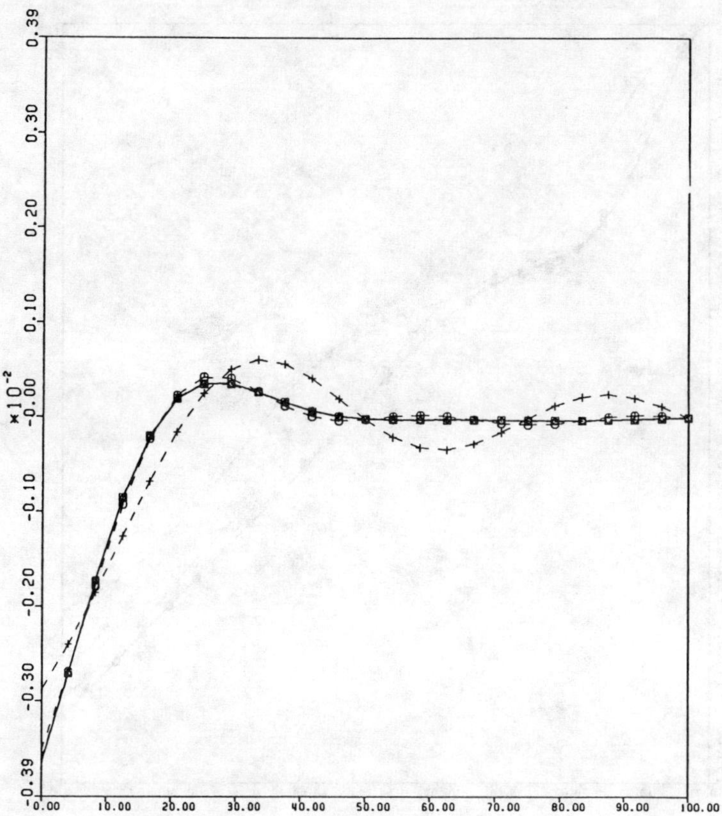

FIG. 3.6(a). *Functional control gain component* ϕ''_{fn}. *Damping coefficient* $c_0 = 10^{-4}$; *control weighting* $R = .05$; *number of flexible modes* $n_m = 4, 6, 8, 10, 12$. *Modal Approximation.*

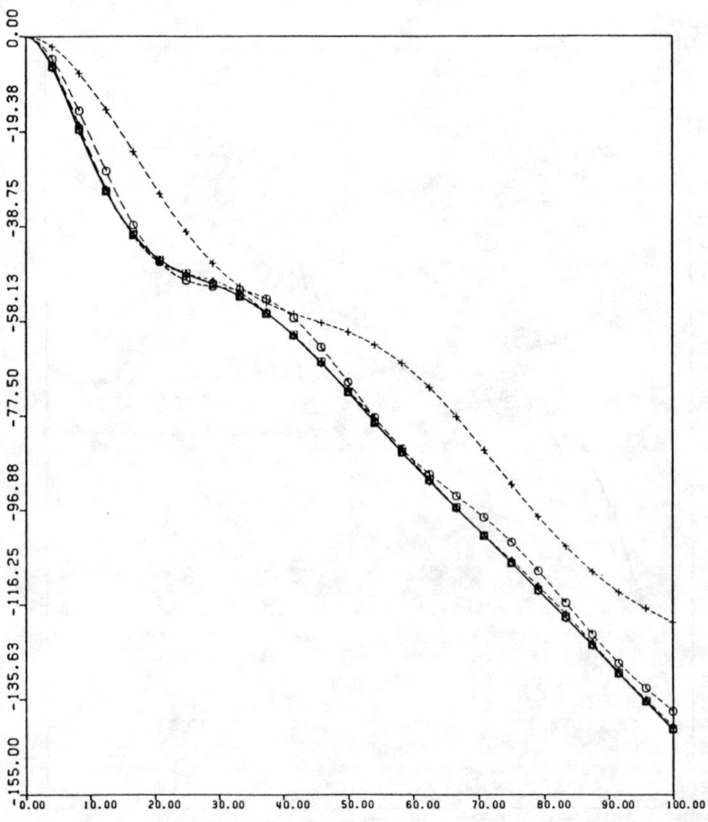

FIG. 3.6(b). *Functional control gain component* ϕ''_{fn}. *Damping coefficient* $c_0 = 10^{-4}$; *control weighting* $R = .05$; *number of flexible modes* $n_m = 4, 6, 8, 10, 12$. *Modal Approximation.*

It is important to note that, for $n_e = n_m$, the B-spline approximation and the natural mode approximation have the same number of degrees of freedom ($n = n_e+1$), while the Hermite spline approximation has $n = 2n_e+1$ degrees of freedom. Both spline approximations have converged for ten elements and the modal approximation has converged for ten flexible modes. The graphs show that the gains for eight elements with Hermite splines are virtually identical to the gains for eight flexible modes (and the rigid-body mode) and only slightly better than the gains for eight elements with B-splines. We conclude that, as basis vectors for approximating the optimal control law, the natural system mode shapes are the best, followed closely by the B-splines, and that the Hermite splines are a distant third, requiring roughly twice as many degrees of freedom for convergence as the other two schemes.

3.6. Approximation of the Infinite-Dimensional Estimator and Compensator for the Example.

As discussed in §3.3.3, the optimal closed-loop system is formed with the optimal infinite-dimensional compensator, which consists of the optimal control law for the distributed model of the structure applied to the output of the optimal infinite-dimensional state estimator. The optimal control law is the limit of the approximating finite-dimensional control laws in §3.5. In this section, we approximate the infinite-dimensional estimator as in §3.4.3 and then apply the approximating control laws in §3.5 to the output of the approximating finite-dimensional estimators to produce finite-dimensional compensators that approximate the optimal compensator.

3.6.1. The Estimator Problem. The single sensor measures the rigid-body angle θ, and we assume that this measurement has zero-mean Gaussian white noise with variance $\widehat{R} = 10^{-4}$. Also, we model a disturbance on the right side of (3.1) that is a zero-mean Gaussian white noise process distributed uniformly over the beam and having concentrated components acting on the hub and tip mass. For this disturbance, the covariance operator $\widehat{Q}$ in (3.33) is

$$\widehat{Q} = \begin{bmatrix} 0 & 0 \\ 0 & I \end{bmatrix} : V \times H \to V \times H. \tag{3.79}$$

We construct the approximating estimators as in §3.4.3. The gain for the nth estimator is given by (3.63) with the solution to the Riccati matrix equation (3.64). For the rigid-body measurement, the matrix C^n is

$$C^n = [1 \ 0 \ 0 \ 0 \ \cdots]. \tag{3.80}$$

According to (3.61), the matrix Q_n is

$$\tilde{\widehat{Q}}^n = \begin{bmatrix} 0 & 0 \\ 0 & M^{-n} \end{bmatrix}, \tag{3.81}$$

since W^n is the matrix in (3.48). (As always, M^{-n} is the inverse of the mass matrix.)

To see the convergence of the approximating estimator gains, we compute the approximating functional estimator gains according to (3.66) and (3.67). Like the functional control gains, the functional estimator gains have the form

$$\hat{f} = (\alpha_f, \phi_f, \beta_f) \in V, \tag{3.82a}$$

$$\hat{g} = (\alpha_g, \phi_g, \beta_g) \in H, \tag{3.82b}$$

and the corresponding approximations have the form

$$\hat{f}_n = (\alpha_{fn}, \phi_{gn}, \beta_{gn}), \tag{3.83a}$$

$$\hat{g}_n = (\alpha_{gn}, \phi_{gn}, \beta_{gn}). \tag{3.83b}$$

As in the control problem, our convergence theory establishes only L_2-convergence for ϕ''_{fn} and ϕ_{gn}, but the numerical results show uniform convergence on $[0,\ell]$. Since $\phi_{fn}(0) = \phi'_{fn}(0) = 0$, the convergence of β''_{fn} implies the convergence of $\beta_{fn} = \phi_{fn}(\ell)$. Thus, β_f is not an independent piece of information about the estimator gains while, as far as our convergence results go, β_g is. We maintain the analogy with the control problem and list only β_{gn} in the subsequent tables.

For the Hermite spline approximation, Figs. 3.6a and 3.6b show ϕ''_{fn} and ϕ_{gn}, Table 3.5 lists the scalars α_{fn}, α_{gn}, and β_{gn}. Figures 3.7a and 3.7b and Table 3.6 show the B-spline results. As in the control problem, all three approximations to the gains converge at about the same rate with n. We show only the two spline approximations, since the modal approximation looks almost identical to the B-spline approximation. As in the control problem, ϕ''_{fn} is continuous for both the B-splines and the modes but discontinuous at the nodes for the Hermite splines.

3.6.2. Approximation of the Optimal Compensator. For each of the three approximation schemes, we have constructed the finite-dimensional compensator in Fig. 3.3 for $n = 5, 7, 9, 11$; i.e., 4, 6, 8, and 10 elements for the splines and 4, 6, 8, and 10 natural mode shapes. Recall that (3.69) gives the transfer function of the nth compensator. Figures 3.9–3.11 show the frequency responses (bode plots) of these compensators. The first graph in each figure contains the gain (magnitude) plots for all four orders of approximation, while the phase plot represents only highest-order approximation.

The finite-dimensional compensators for each approximation scheme converge to the same infinite-dimensional compensator as n increases, and this is consistent with the convergence of the functional control and estimator gains. However, we should point out a slight difference among the bode plots for

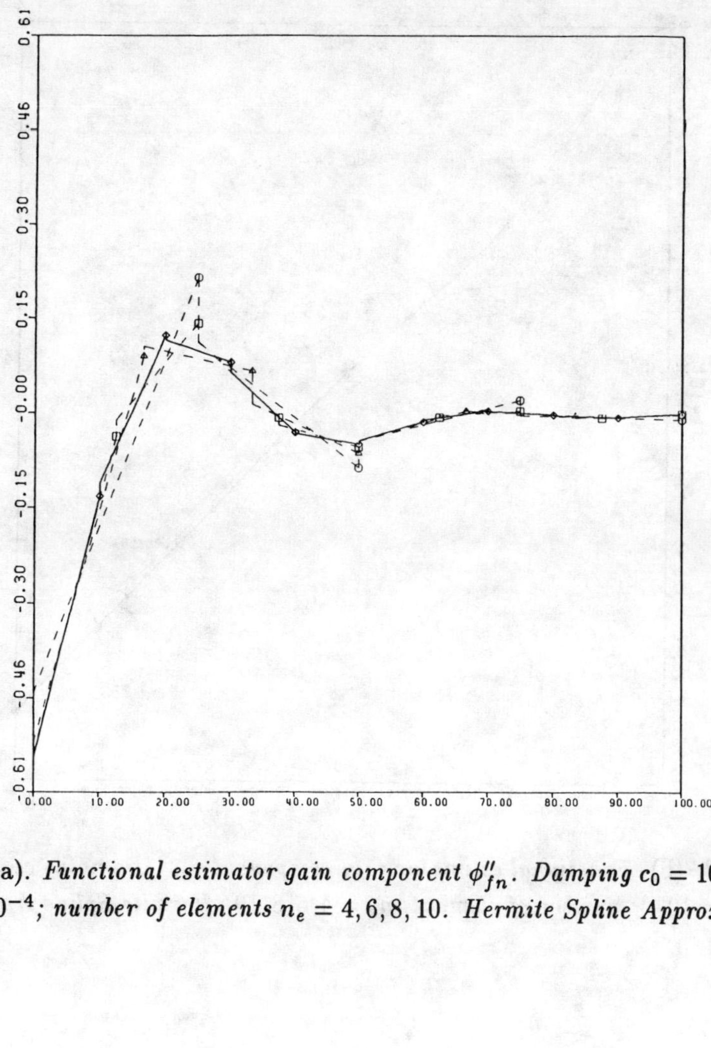

FIG. 3.7(a). *Functional estimator gain component* ϕ''_{fn}. *Damping* $c_0 = 10^{-4}$; *estimator* $\widehat{R} = 10^{-4}$; *number of elements* $n_e = 4, 6, 8, 10$. *Hermite Spline Approximation.*

TABLE 3.5

Scalar components of functional estimator gains. Damping coefficient $c_0 = 10^{-4}$; *estimator* $\widehat{R} = 10^{-4}$; *number of elements* $n_e = 4, 6, 8, 10$. *Hermite Spline Approximation.*

n_e	α_{fn}	α_{gn}	β_{gn}
4	5.3195	14.149	−1495.7
6	5.3567	14.347	−1517.5
8	5.3611	14.371	−2520.1
10	5.3623	14.377	−1520.8

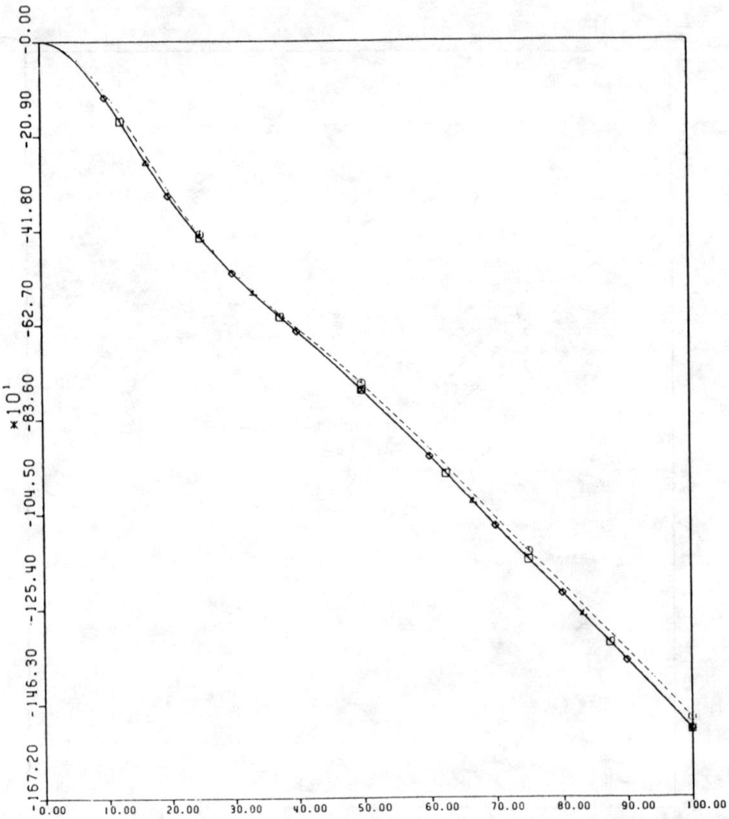

FIG. 3.7(b). *Functional estimator gain component* ϕ_{gn}. *Damping* $c_0 = 10^{-4}$; *estimator* $\widehat{R} = 10^{-4}$; *number of elements* $n_e = 4, 6, 8, 10$. *Hermite Spline Approximation.*

TABLE 3.6

Scalar components of functional estimator gains. Damping coefficient $c_0 = 10^{-4}$; *estimator* $\widehat{R} = 10^{-4}$; *number of elements* $n_e = 4, 6, 8, 10, 12$. *B-Spline Approximation.*

n_e	α_{fn}	α_{gn}	β_{gn}
4	5.2749	13.912	−1469.3
6	5.3538	14.332	−1515.8
8	5.3606	14.368	−1519.8
10	5.3621	14.376	−1520.7
12	5.3626	14.379	−1521.0

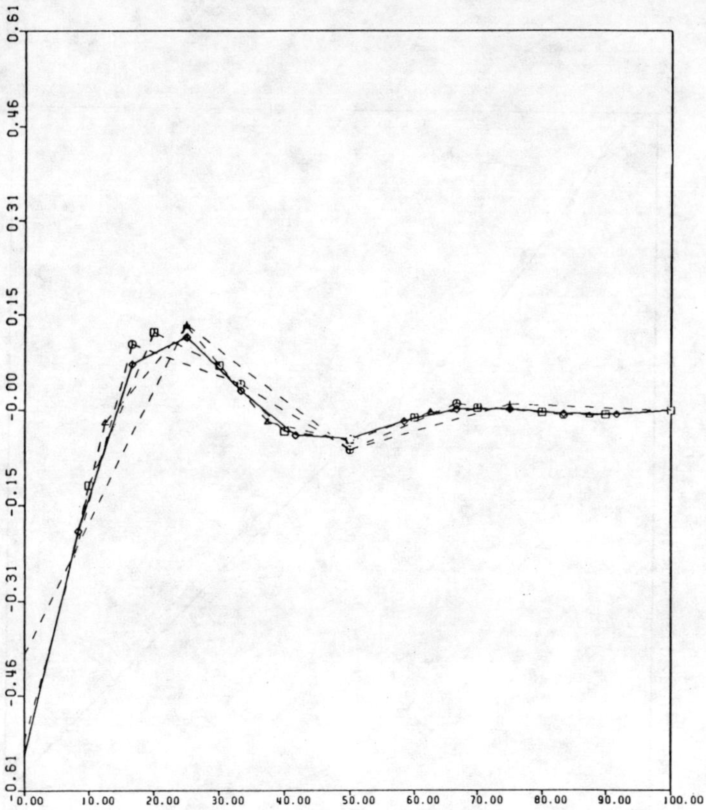

FIG. 3.8(a). *Functional estimator gain component* ϕ''_{fn}. *Damping* $c_0 = 10^{-4}$; *estimator* $\widehat{R} = 10^{-4}$; *number of elements* $n_e = 4, 6, 8, 10, 12$. *B-Spline Approximation.*

$n_e = 10$ in Figs. 3.9 and 3.10 and $n_m = 10$ in Fig. 3.11 (the solid curve in each plot). The peaks at 32 rad/s in the three figures have different heights. Examination of closed-loop eigenvalues and of the bode plots based on splines for 12 elements show that the solid curve in Fig. 3.11 is the bode plot for the infinite-dimensional compensator and therefore the limit for all three approximation schemes. Hence, the compensators based on Hermite splines and B-splines have not quite converged for 10 elements. While we have computed higher-order spline-based compensators than those shown here, we have chosen the approximation orders here to best illustrate the differences among the three approximation schemes.

3.7. The Closed-Loop Eigenstructure.

As we said in (3.41), the spectrum of the optimal closed-loop system is the union of the spectrum of $(A - BF)$ and the spectrum of $(A - \widehat{F}C)$. If the damping operator D_0 in (3.1) is bounded relative to A_0^{μ} for some $\mu < 1$, then A has compact resolvent. In this case, $(A - BF)$ and $(A - \widehat{F}C)$ also have compact resolvent because BF and $\widehat{F}C$ are bounded, so that the closed-loop spectrum consists only of isolated eigenvalues, each with finite multiplicity. If,

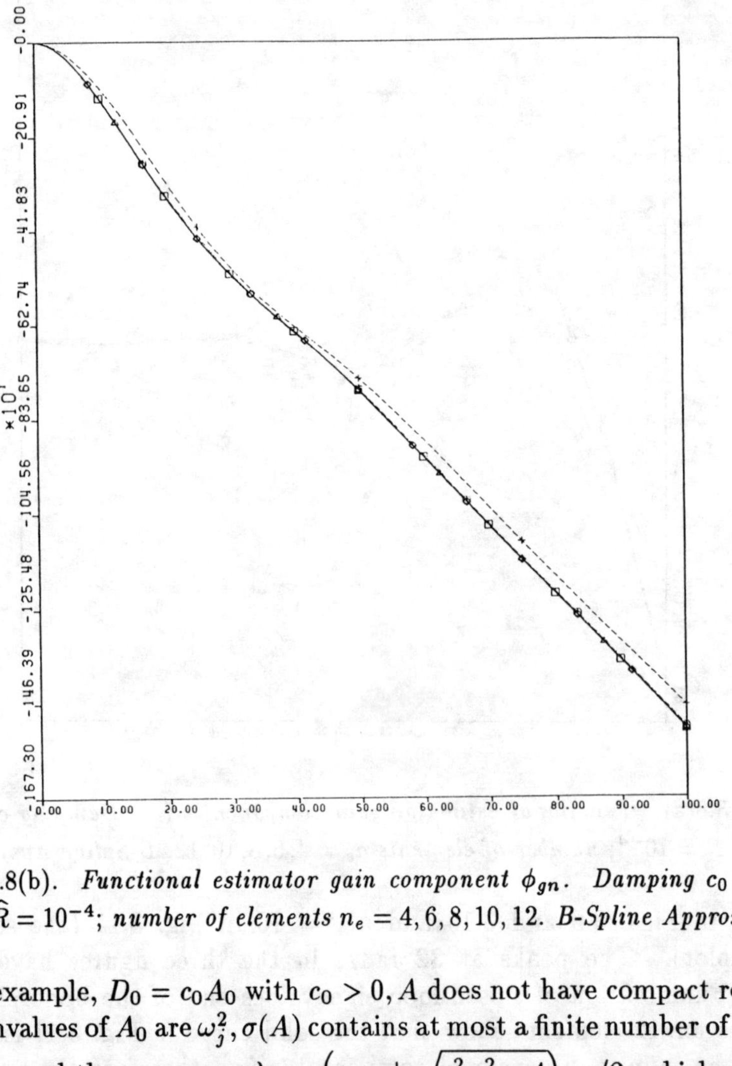

FIG. 3.8(b). *Functional estimator gain component ϕ_{gn}. Damping* $c_0 = 10^{-4}$; *estimator* $\widehat{R} = 10^{-4}$; *number of elements* $n_e = 4, 6, 8, 10, 12$. *B-Spline Approximation.*

as in our example, $D_0 = c_0 A_0$ with $c_0 > 0, A$ does not have compact resolvent; if the eigenvalues of A_0 are $\omega_j^2, \sigma(A)$ contains at most a finite number of complex eigenvalues and the sequences $\lambda_j = \left(-c_0 \pm \sqrt{c_0^2\omega_j^2 - 4}\right)\omega_j/2$, which approach $-\infty$ and the continuous spectrum $\{-1/c_0\}$ as j increases. When $D_0 = c_0 A_0^\mu$ for any $0 \le \mu \le 1$, the eigenvectors of A are the same as those when $D_0 = 0$, and hence are complete in E.

For the remainder of this discussion, let us assume that, as in our example, the spectrum of A (ie., the open-loop spectrum) consists of isolated eigenvalues with finite multiplicity and possibly a finite number of limit points, and that the eigenvectors are complete in E. (This is the case in our example.) Then, since BF and $\widehat{F}C$ are bounded with finite rank, it follows from applying standard perturbation results [11, pp. 208–214] that the asymptotic properties of both $\sigma(A - BF)$ and $\sigma(A - \widehat{F}C)$ are identical to those of $\sigma(A)$. This means that, beyond some number of eigenvalues, or modes, the eigenvalues of $A - BF$ are virtually identical to those of A because the optimal compensator essentially controls only a finite number of modes. It also means that, although the

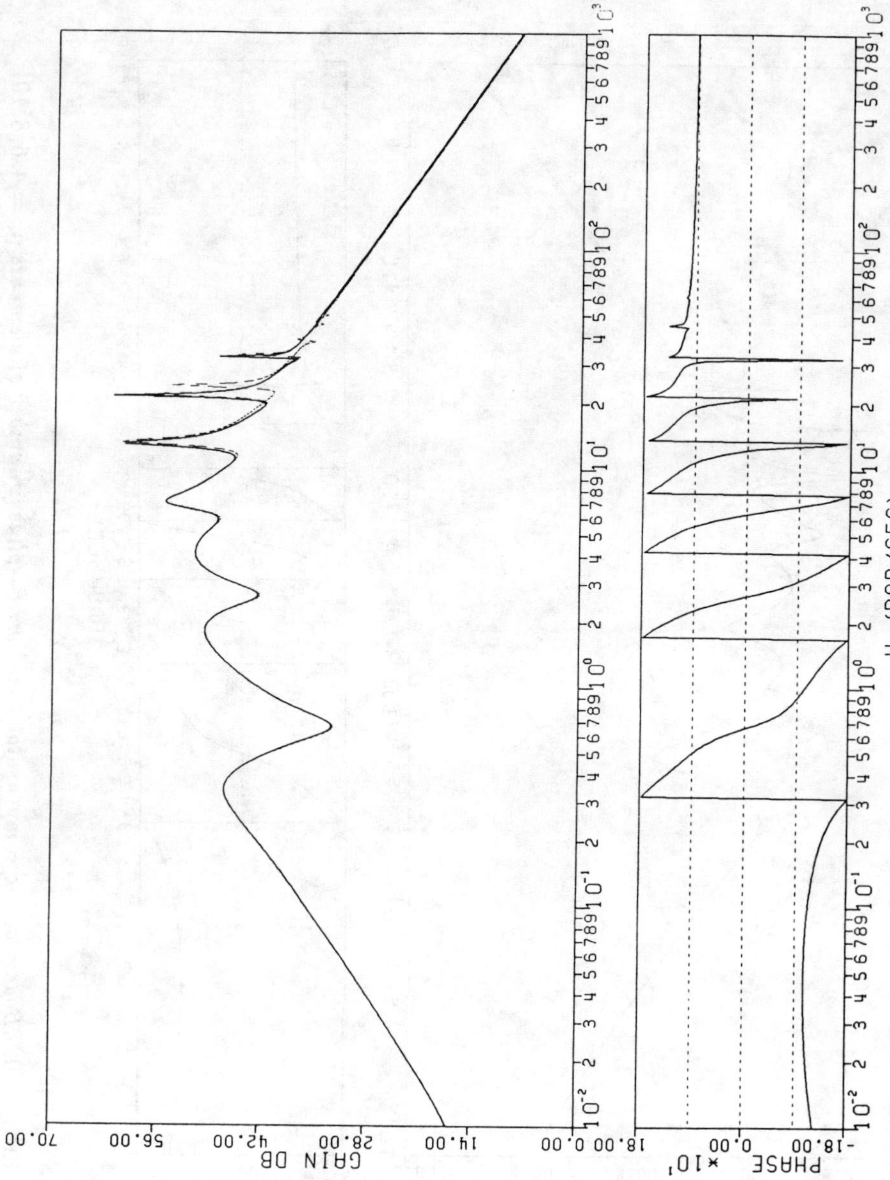

FIG. 3.9. *Bode plot for compensator based on Hermite splines; number of elements* $n_e = 4, 6, 8, 10$.

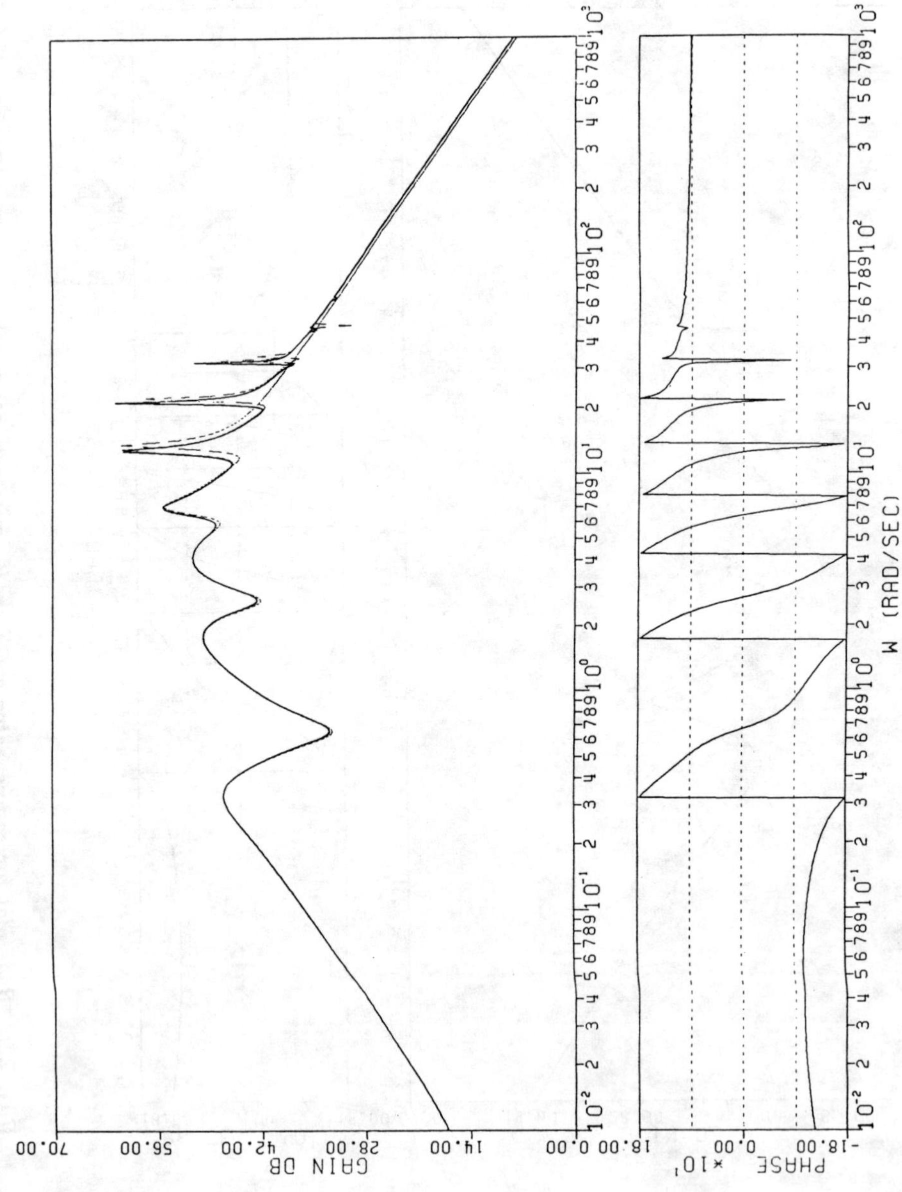

FIG. 3.10. *Bode plot for compensator based on B-splines; number of elements* $n_e = 4, 6, 8, 10$.

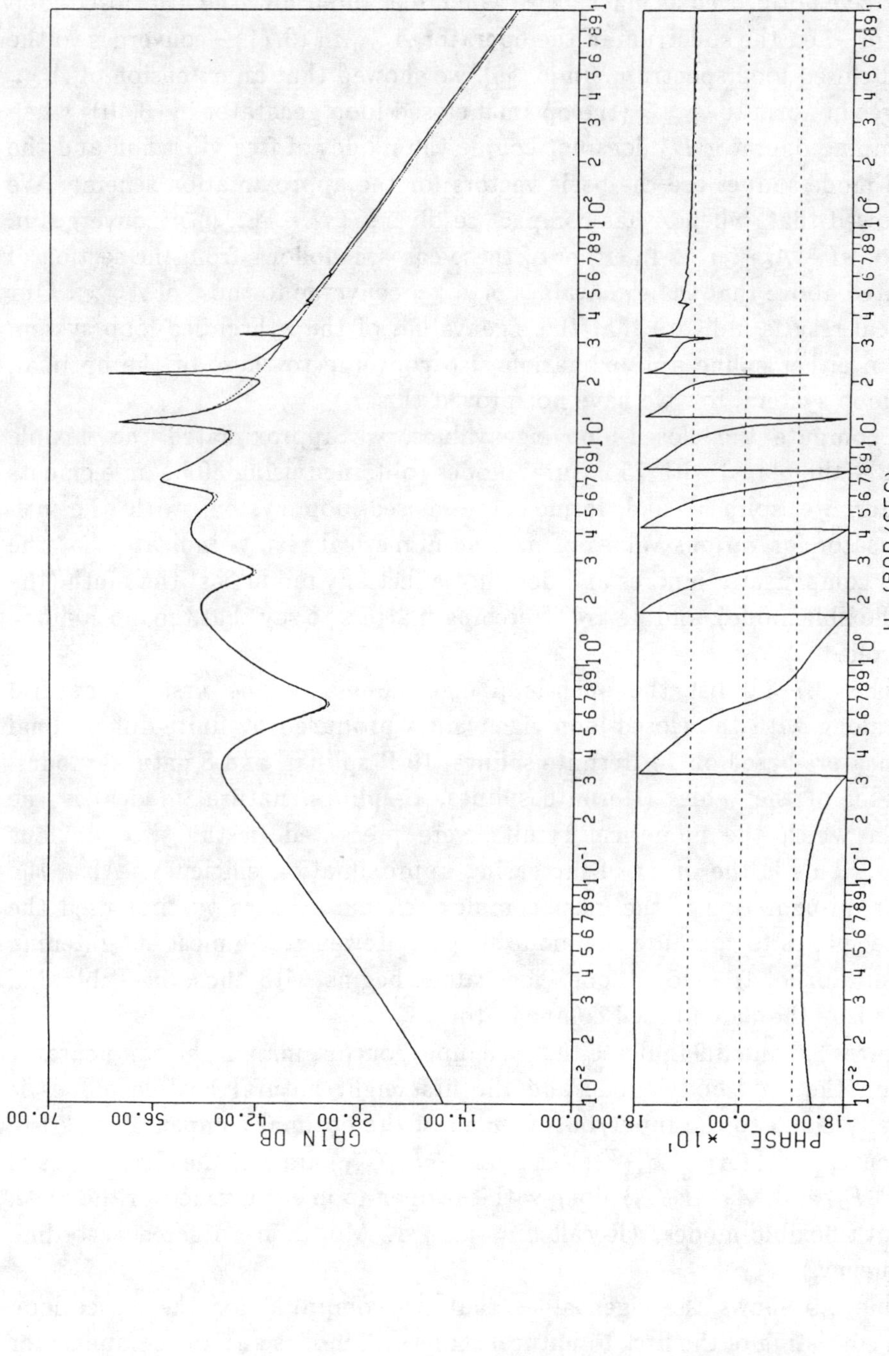

FIG. 3.11. *Bode plot for compensator based on natural modes; number of flexible modes* $n_m = 4, 6, 8, 10$.

optimal estimator contains copies of all the modes of the structure, essentially it observes only a finite number of modes and feeds virtually no sensor data into the copies of the rest. Of course, the infinite number of inactive estimator modes should be truncated before implementation.

We have not proved as much as we would like about how the nth closed-loop spectrum—i.e., the spectrum of the operator $A_{\infty,n}$ in (3.71)—converges to the optimal closed-loop spectrum. In [6, §9], we showed that an extension of $A_{\infty,n}$ converges in norm to $A_{\infty,\infty}$ (the optimal closed-loop generator in (3.40)) when the damping operator D_0 does not couple the modes of free vibration and the natural mode shapes are the basis vectors for the approximation scheme. We also showed that, when A has compact resolvent, $(sI - A_{\infty,n})^{-1}$ converges in norm to $(sI - A_{\infty,\infty})^{-1}$. In either of these cases, it follows from the section of Kato cited above that the eigenvalues of $A_{\infty,n}$ converge to those of $A_{\infty,\infty}$. Our numerical results indicate that the eigenvalues of the nth closed-loop system based on either spline approximation also converge to those of the optimal, closed-loop system, but we have not proved this.

To compute the closed-loop eigenvalues, we approximated the flexible structure (the plant) with 15 natural modes (obtained using 30 beam elements with Hermite splines) and formed the closed-loop system with the nth compensator for various values of n. The numerical results indicate that the optimal compensator ignores and does not affect any mode past the ninth (the eighth flexible mode), and we are stretching matters to say that the ninth mode is controlled.

Tables 3.7–3.9 list the open-loop eigenvalues for the first 15 natural modes along with the closed-loop eigenvalues produced by finite-dimensional compensators based on 10 Hermite splines, 10 B-splines, and 8 natural modes. The order of the tables (Hermite splines, B-splines, natural modes) is the order in which the numerical results were presented in the two previous sections. This is the order of increasing approximation efficiency. Also, the modal approximation is the most complex scheme because we first used the Hermite splines to compute the mode shapes. However, the most enlightening interpretation of the closed-loop eigenvalues begins with those in Table 3.9, produced by the mode-based compensator.

Figures 3.6 and 3.9 indicate that the input/output map of the compensator based on the rigid-body mode and the first eight natural flexible modes is essentially equal to the input/output map of the optimal compensator. Also, the eigenvalues of $(A_{11} - B_{11}F_{11})$ and $(A_{11} - \widehat{F}_{11}C_{11})$ are just the eigenvalues of $(A_9 - B_9F_9)$ and $(A_9 - \widehat{F}_9C_9)$ along with the open-loop eigenvalues for the ninth and tenth flexible modes. (Recall $n = n_m + 1$. Modes are ordered according to frequency.)

Table 3.9 shows the eigenvalues that we computed for the closed-loop system consisting of the first 15 natural structural modes and the compensator for $n = 9$. These eigenvalues are equal to the eigenvalues of $(A_9 - B_9F_9)$ and $(A_9 - \widehat{F}_9C_9)$ and the open-loop eigenvalues of modes 10 through 15.

The six pairs of unshifted eigenvalues represent those modes in the 15-mode plant model that the compensator does not affect. We list such modes in the $(A - BF)$ column. In each of the three groups of eigenvalues in each table, the first two eigenvalues correspond to the rigid-body mode and the rest correspond to the flexible modes. The spectrum of the closed-loop system consisting of the entire flexible structure and this finite-dimensional compensator contains the closed-loop eigenvalues in Table 3.9, the remaining open-loop eigenvalues (for modes 16 and above) and the limit point -10^4 (the one point continuous spectrum). We would list the eigenvalues for the remaining uncontrolled modes in the $(A - BF)$ column.

The optimal infinite-dimensional compensator consists of the compensator for $n = 9$ along with copies of the remaining uncontrolled modes. In the compensator, these modes are inactive in that they do not contribute to the input/output map of the compensator. Hence, they should not be included in the implemented compensator. Probably, the ninth mode of the compensator should be truncated, also.

The closed-loop eigenvalues that would result from the compensator for smaller values of n would not equal the eigenvalues of $(A_n - B_n F_n)$ and $(A_n - \widehat{F}_n C_n)$ and the open-loop eigenvalues of modes greater than n because of the coupling between modes modeled in the compensator with modes not modeled in the compensator. This is the coupling known as spillover (Balas' term [2], [3]). Spillover is eliminated when the finite-dimensional compensators converge to the infinite-dimensional compensator. See the discussion in [6, §9.2].

Table 3.7 lists the eigenvalues for the closed-loop system consisting of the 15-mode model of the structure and the compensator based on 10 elements with Hermite splines. For this compensator, $n = 2 * 10 + 1 = 21$, so that the estimator contains approximations to the first 21 modes of the structure. Most of these approximations have not converged to the true natural modes, as can be seen from the imaginary parts of the eigenvalues under $A - \widehat{F}C$, which converge from above to slight perturbations of the open-loop frequencies.

The last twelve of the twenty-one modes in the Hermite spline compensator are inactive as far as the input/output map of the compensator. They do not couple with the rest of the compensator or any modes in the structure, and they could be truncated without affecting the closed-loop response. The eigenvalues of the inactive compensator modes are equal to the corresponding open-loop eigenvalues of the ten-element approximation to the structure. The magnitude of real parts are larger than those of the corresponding true open-loop eigenvalues because the damping is proportional to the stiffness and the frequencies of the higher modes of the approximation are greater than the true frequencies. Since the inactive active compensator modes reside in the nth estimator, we list the eigenvalues under $(A - \widehat{F}C)$.

The closed-loop eigenvalues in Table 3.7 corresponding to modes 5, 6, and 7 are not equal to the corresponding eigenvalues of $(A_{21} - B_{21}F_{21})$ and

$(A_{21} - \widehat{F}_{21}C_{21})$ for the Hermite splines, nor are they equal to the corresponding eigenvalues in Table 3.9. (For each of modes 5, 6, and 7, we listed one pair of closed-loop eigenvalues under $(A-BF)$ and one pair under $(A-\widehat{F}C)$, according to the eigenvalues in Table 3.9 that each pair most closely approximates.) This means that the Hermite-spline compensator for ten elements is not as close to convergence as Figs. 3.4, 3.7, and 3.10 seem to suggest, but it should not be surprising in view of the difference between the peak at mode 7 (32 rad/s) in Fig. 3.9 and the corresponding peak in Fig. 3.11. Still, the closed-loop response produced by the ten-element Hermite-spline compensator is quite close to that produced by the mode-based compensator, except possibly for the sixth mode.

Although the mode-based compensator for $n = 9$ is essentially identical, as far as input/output and the closed-loop system that it produces, to the optimal infinite-dimensional compensator while the Hermite-spline compensator for $n = 21$ is not quite there, the mode-based compensator should be compared to the compensator obtained by truncating the 12 inactive modes of the Hermite-spline compensator. Then implementable compensators of equal dimension $(2 * 9 = 18)$ are being compared, and the two compensators differ only slightly. After convergence and truncation of the inactive compensator modes, the Hermite splines do not lead to a larger compensator than the mode-based compensator. The Hermite splines are inefficient for compensator design because they require so many degrees of freedom to approximate the active compensator modes, thereby necessitating the solution of Riccati equations more than twice the dimension of those solved to design the mode-based compensator.

Table 3.8 lists the closed-loop eigenvalues produced by the compensator based on ten elements with B-splines. The only significant difference between the B-spline compensator and the Hermite-spline compensator is that the B-spline compensator has ten fewer inactive modes. Still, there are at least two that can be truncated. The eigenvalues indicate that, as far as the flexible structure is concerned, both spline-based compensators produce virtually the same closed-loop system. Although the Riccati equations that must be solved before the B-spline compensator converges to the optimal compensator are still larger than the Riccati equations that must be solved when the natural modes are used, the B-splines avoid the preliminary finite element approximation of the natural mode shapes at a relatively small price in terms of the dimension of the Riccati equations.

3.8. Conclusions.

As predicted by the results in [6], the finite-dimensional compensators designed by each of our three approximation schemes converge to the same infinite-dimensional compensator as the order of approximation increases. The rates of convergence, though, are quite different. All of our comparisons show that, of the three approximation schemes in this paper, the scheme using natural mode shapes as basis functions is the best in terms of the number of degrees

TABLE 3.7

Closed-loop eigenvalues with compensator based on 10 Hermite splines.

Open-Loop		$A - BF$		$A - \widehat{F}C$	
REAL	IMAG	REAL	IMAG	REAL	IMAG
0.0	0.0	−0.03305	0.0	−0.27579	0.33295
0.0	0.0	−0.00799	0.0	−0.27579	−0.33295
−0.00005	0.96522	−0.11498	0.96522	−1.07059	1.52394
−0.00052	3.21201	−0.17909	3.21192	−0.83359	3.27848
−0.00242	6.96055	−0.15092	6.95926	−0.33913	6.97026
−0.00840	12.96107	−0.09170	12.94415	−0.14130	12.98792
−0.02280	21.35301	−0.03957	21.33965	−0.09902	21.40999
−0.05149	32.08958	−0.05132	32.08852	−0.08468	32.23854
−0.10190	45.14356	−0.10176	45.14358	−0.11477	45.54628
−0.18305	60.50577	−0.18300	60.50578	−0.19281	61.43464
−0.30559	78.17711	−0.30557	78.17711	−0.32191	80.02459
−0.48184	98.16564	−0.48183	98.16564	−0.51073	100.99211
−0.72587	120.48589	−0.72586	120.48589	−0.87879	132.55647
−1.05361	145.15908	−1.05361	145.15908	−1.28294	160.16850
−1.48299	172.21366	−1.48299	172.21366	−1.89848	194.84366
−2.03407	201.68604	−2.03407	201.68604	−2.78066	235.80614
				−4.03103	283.90815
				−5.78013	339.95430
				−8.15257	403.71367
				−11.13836	471.85059
				−14.29842	534.56892
				−16.50298	574.27103

TABLE 3.8

Closed-loop eigenvalues with compensator based on 10 B-*splines.*

Open-Loop		$A-BF$		$A-\widehat{F}C$	
REAL	IMAG	REAL	IMAG	REAL	IMAG
0.0	0.0	−0.03305	0.0	−0.27579	0.33295
0.0	0.0	−0.00799	0.0	−0.27579	−0.33295
−0.00005	0.96522	−0.11498	0.96522	−1.07059	1.52394
−0.00052	3.21201	−0.17909	3.21191	−0.83360	3.27849
−0.00242	6.96055	−0.15086	6.95904	−0.33920	6.97068
−0.00840	12.96107	−0.08840	12.94195	−0.14463	12.99319
−0.02280	21.35301	−0.03449	21.34000	−0.10421	21.43490
−0.05149	32.08958	−0.05114	32.08911	−0.08539	32.37767
−0.10190	45.14356	−0.10176	45.14360	−0.11752	46.12272
−0.18305	60.50577	−0.18300	60.50578	−0.20434	63.26800
−0.30559	78.17711	−0.30557	78.17711	−0.35598	84.17569
−0.48184	98.16564	−0.48183	98.16564	−0.55429	105.24019
−0.72587	120.48589	−0.72586	120.48589		
−1.05361	145.15908	−1.05361	145.15908		
−1.48299	172.21366	−1.48299	172.21366		
−2.03407	201.68604	−2.03407	201.68604		

of freedom required for convergence. However, the B-splines are almost as good as the modes. The Hermite splines require roughly twice the number of degrees of freedom for convergence as do the modes. The only disadvantage of the modal approximation is that it requires a preliminary finite element approximation to obtain the mode shapes and frequencies.

The three compensators compared in §3.7 have almost identical transfer functions, and in this sense approximate the optimal infinite-dimensional compensator. However, they have very different structures because of the inactive modes in the spline-based compensators. Direct modal truncation of the compensator, or a method like balanced realizations, if necessary [9], [14], will reduce the spline-based compensators to the same dimension as the mode-based compensator, but the fact that this second stage of approximation is either unnecessary or much easier for the smaller mode-based compensator is another significant advantage of the modal approximation.

In view of the results here on approximating the optimal compensator and the results in [15] on approximating the natural modes of a structure consisting of a flexible beam with tip bodies, it is hard to see any reason for using Hermite splines at all in compensator design for flexible structures with beams. The best approach should be either to use B-splines (or perhaps some even smoother

TABLE 3.9
Closed-loop eigenvalues with compensator based on 8 modes.

Open-Loop		$A-BF$		$A-\hat{F}C$	
REAL	IMAG	REAL	IMAG	REAL	IMAG
0.0	0.0	−0.03305	0.0	−0.27580	0.33295
0.0	0.0	−0.00799	0.0	−0.27580	−0.33295
−0.00005	0.96522	−0.11498	0.96522	−1.07061	1.52391
−0.00052	3.21201	−0.17914	3.21198	−0.83348	3.27829
−0.00242	6.96055	−0.15116	6.96055	−0.33889	6.96771
−0.00840	12.96107	−0.10511	12.96111	−0.12795	12.96164
−0.02280	21.35301	−0.07855	21.35300	−0.06015	21.35310
−0.05149	32.08958	−0.07706	32.08958	−0.05873	32.08959
−0.10190	45.14356	−0.11176	45.14356	−0.10316	45.14357
−0.18305	60.50577	−0.18697	60.50577	−0.18332	60.50577
−0.30559	78.17711	−0.30557	78.17711		
−0.48184	98.16564	−0.48183	98.16564		
−0.72587	120.48589	−0.72586	120.48589		
−1.05361	145.15908	−1.05361	145.15908		
−1.48299	172.21366	−1.48299	172.21366		
−2.03407	201.68604	−2.03407	201.68604		

basis functions) for the approximating control problems, or to use B-splines to compute the natural modes of the structure and then use the modes shapes as the basis functions in the approximating control problems.

References.

[1] A. V. Balakrishnan, *Applied Functional Analysis*, Second ed., Springer-Verlag, New York, 1981.

[2] M. J. Balas, *Feedback Control of Flexible Systems*, IEEE Trans. AC, AC–23, pp. 673–679, 1978.

[3] M. J. Balas, *Modal control of certain flexible systems*, SIAM J. Control Optim., 16(1978) pp. 450–462.

[4] R. F. Curtain and A. J. Pritchard, *Infinite Dimensional Linear Systems Theory*, Springer-Verlag, New York, 1978.

[5] J. S. Gibson, *An analysis of optimal modal regulation: Convergence and stability*, SIAM J. Control Optim., 19 (1981), pp. 686–707.

[6] J. S. Gibson and A. Adamian, *Approximation theory for Linear-Quadratic-Gaussian control of flexible structures*, SIAM J. Control Optim., 29 (1991), pp. 1–37.

[7] J. S. Gibson and D. L. Mingori, *Approximation and Compensator Order Determination in Optimal Control of Flexible Structures*, Proc. IFAC Workshop on Model Error Concepts and Compensation, Boston, MA, June 1985.

[8] J. S. Gibson, D. L. Mingori, A. Adamian, and F. Jabbari, *Approximation of Optimal Infinite Dimensional Compensators for Flexible Structures*, JPL Workshop on Identification and Control of Space Structures, San Diego, CA, June 1984.

[9] J. S. Gibson, D. L. Mingori, et al., HR Textron Report #956541 – Final, to JPL, September 1984.

[10] J. S. Gibson and M. Navid, *Approximate Solution of Riccati Algebraic Equations in Optimal Control and Estimation of Hyperbolic Systems*, International Symposium on Mathematical Theory of Networks and Systems, Vol. 4, Santa Monica, CA, August 1981.

[11] T. Kato, *Perturbation Theory for Linear Operators*. Second ed., Springer-Verlag, New York, 1984.

[12] H. Kwakernaak and R. Sivan, *Linear Optimal Control Systems*, Wiley Interscience, New York, 1972.

[13] D. L. Mingori, J. S. Gibson, P. Blelloch, and A. Adamian, *Control of a Flexible Space Antenna: A Finite Dimensional Perspective Based on Distributed Parameter Theory*, JPL Workshop on Identification and Control of Space Structures, San Diego, CA, June 1985.

[14] B. C. Moore, *Principal Component Analysis in Linear Systems: Controllability, Observability, and Model Reduction*, IEEE Trans. Automat. Contr., AC-26 (1981), pp. 17–32.

[15] I. G. Rosen, *Spline-based Rayleigh-Ritz methods for the approximation of the natural modes of vibration for flexible beams with tip bodies*, Quart. Appl. Math., 44 (1986), pp. 169–185.

[16] M. H. Schultz, *Spline Analysis*. Prentice-Hall, Englewood Cliffs, N.J., 1973.

[17] R. E. Showalter, *Hilbert Space Methods for Partial Differential Equations*, Pitman, London, 1977.

[18] G. Strang and G. J. Fix, *An Analysis of the Finite Element Method*, Prentice-Hall, Englewood Cliffs, N.J., 1973.

Chapter 4

On Mathematical Models for the Elastic Beam with Frequency-Proportional Damping

David L. Russell

4.1. Introduction to Elementary Theories of Elastic Beam Motion.

The less-than-dashing, rather pedestrian term "beam" has a guaranteed soporific effect on all but the more dedicated devotees of the science of structural mechanics. But it refers, of course, to the indispensable *linear structural elements* (those seeking new terminology, please take note!) without which many complex constructions would be infeasible. Consequently, it is not surprising that mathematical models for the elastic beam have an ancient and honorable pedigree attested to by the attentions of some of the most honored historical and current figures of applied mathematics.

The *Euler–Bernoulli* equation for the motion of thin elastic beams is at least two hundred years old. Denoting the mass density, per unit length, by $\rho(x)$ and the second moment of the modulus of elasticity about the elastic axis (about which the first moment of the modulus of elasticity vanishes) by $EI(x)$, $0 \le x \le L$, it is assumed that the energy associated with motion in the x, w plane, in which the elastic axis is given by $w = 0$, can be adequately represented by

$$(4.1) \qquad \mathcal{E}\left(w, \frac{\partial w}{\partial t}\right) = \int_0^L \left[\rho\left(\frac{\partial w}{\partial t}\right)^2 + EI\left(\frac{\partial^2 w}{\partial x^2}\right)^2\right] dx.$$

If it is assumed that no work is done on the beam, either by externally applied forces or by internal dissipative mechanisms, an easy application of the principle of virtual work shows that the motion must be governed by the partial differential equation

$$(4.2) \qquad \rho\frac{\partial^2 w}{\partial t^2} + \frac{\partial^2}{\partial x^2}\left(EI\frac{\partial^2 w}{\partial x^2}\right) = 0$$

together with "natural boundary conditions" (four are required in all) implying that

$$EI\frac{\partial^2 w}{\partial x^2}\frac{\partial^2 w}{\partial t\,\partial x}\bigg|_0^L - \frac{\partial}{\partial x}\left(EI\frac{\partial^2 w}{\partial x^2}\right)\bigg|_0^L = 0.$$

A variety of conditions, corresponding to different kinematic and dynamic assumptions, suffice for this purpose. The "cantilever" configuration, wherein the beam is clamped at one end, say $x = 0$, and left free at the other, $x = L$, is described by

$$W(0,t) = 0, \qquad \frac{\partial w}{\partial x}(0,t) = 0, \tag{4.3}$$

$$EI(L)\frac{\partial^2 w}{\partial x^2}(L,t) = 0, \qquad -\frac{\partial}{\partial x}\left(EI\frac{\partial^2 w}{\partial x^2}\right)\bigg|_{x=L} = 0. \tag{4.4}$$

If a lateral force, φ, and moment, μ, act at $x = L$, then the right-hand sides of the equations in (4.4) are replaced by μ and φ, respectively. We will describe some other cases subsequently.

The *Rayleigh* beam model is a fairly modest modification of the Euler–Bernoulli model, dating back some hundred years or so. To obtain it, the beam element of width δ centered at x is additionally endowed with mass moment of inertia $\delta I_\rho(x)$. The term $I_\rho\left(\partial^2 w/\partial t\,\partial x\right)^2$ is added to the energy integrand of (4.1) and there results the partial differential equation

$$\rho\frac{\partial^2 w}{\partial t^2} - \frac{\partial}{\partial x}\left(I_\rho\frac{\partial^3 w}{\partial t^2 \partial x}\right) + \frac{\partial^2}{\partial x^2}\left(EI\frac{\partial^2 w}{\partial x^2}\right) = 0 \tag{4.5}$$

with corresponding modifications of the boundary conditions which we will not detail at this point.

The *Timoshenko* beam model goes back about fifty years and represents a more radical departure from the simple Euler–Bernoulli theory than does the Rayleigh model. In the Timoshenko model [22] beam elements undergo not only rigid motions, as is assumed in the previous two models, but also undergo a shearing motion so that an originally rectangular w, x cross section is transformed into a parallelogram as shown in Fig. 4.1(b). The total rotation angle of an element is thus

$$\psi(x,t) = \theta(x,t) + \beta(x,t)$$

where for small motions the bending angle θ is given approximately by

$$\theta(x,t) \simeq \frac{\partial w}{\partial x}(x,t)$$

and β is the angle of shear. Consequently,

$$\beta(x,t) \simeq \psi(x,t) - \frac{\partial w}{\partial x}(x,t).$$

The energy expression used is

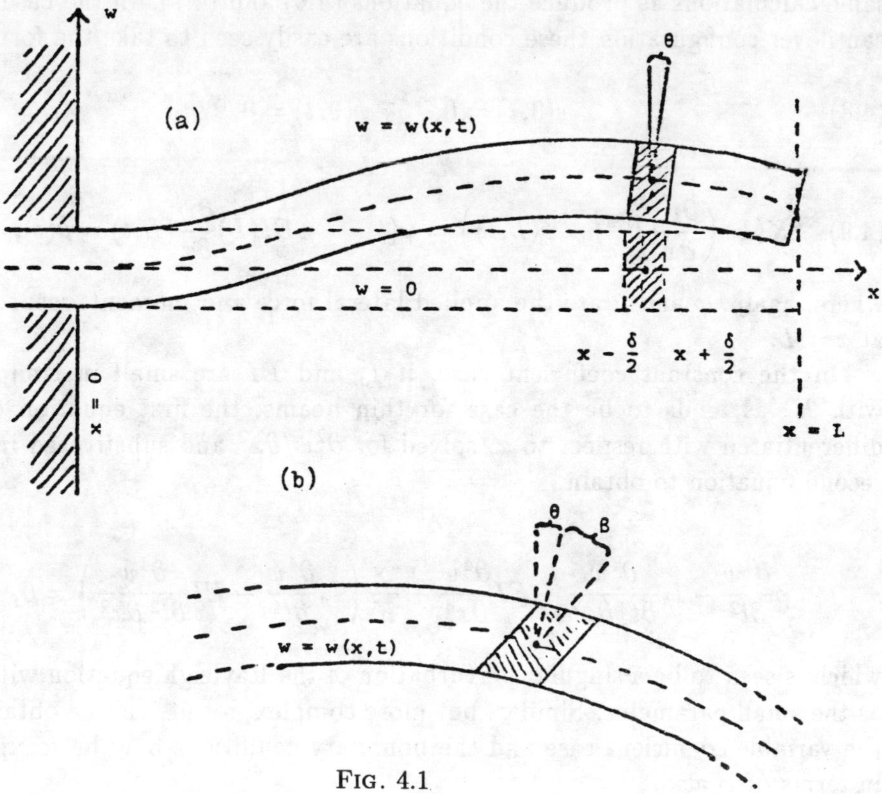

FIG. 4.1

$$\mathcal{E}\left(w, \frac{\partial w}{\partial t}, \psi, \frac{\partial \psi}{\partial t}\right) = \int_0^L \left[\rho\left(\frac{\partial w}{\partial t}\right)^2 + I_\rho\left(\frac{\partial \psi}{\partial t}\right) + K\left(\psi - \frac{\partial w}{\partial x}\right)^2 + EI\left(\frac{\partial \psi}{\partial x}\right)\right] dx,$$

where $K(x)$ is the modulus of elasticity in shear. Again, application of the principle of virtual work leads to the Timoshenko equations

$$\rho\frac{\partial^2 w}{\partial t^2} - \frac{\partial}{\partial x}\left(K\frac{\partial w}{\partial x} - \psi\right) = 0, \tag{4.6}$$

$$I_\rho\frac{\partial^2 \psi}{\partial t^2} - \frac{\partial}{\partial x}\left(EI\frac{\partial \psi}{\partial x}\right) + K\left(\psi - \frac{\partial w}{\partial x}\right) = 0. \tag{4.7}$$

As inspection quickly shows, these equations have the form of two coupled wave equations. In general the wave speeds $(K(x)/\rho(x))^{1/2}$ and $(EI(x)/I_\rho(x))^{1/2}$ are different, resulting in a hyperbolic system with four families of characteristic curves, consisting of two pairs, each pair corresponding to waves of the same speed moving in opposite directions. The boundary conditions come out of the same calculations as produce the equations (4.6) and (4.7). In the case of the cantilever configuration these conditions are easily seen to take the form

$$w(0,t) = 0, \qquad \psi(0,t) = 0, \tag{4.8}$$

$$K(L)\left(\frac{\partial w}{\partial x}(L,t) - \psi(L,t)\right) = \varphi(t), \qquad EI(L)\frac{\partial \psi}{\partial x}(L,t) = \mu(t), \tag{4.9}$$

where, again, φ and μ are the applied lateral force and moment, respectively, at $x = L$.

In the constant coefficient case, if I_ρ and EI are small in comparison with K, as tends to be the case for thin beams, the first equation can be differentiated with respect to x, solved for $\partial^2\psi/\partial x^2$ and substituted into the second equation to obtain

$$\rho\frac{\partial^2 w}{\partial t^2} - I_\rho\frac{\partial^4 w}{\partial t^2\,\partial x^2} + EI\frac{\partial^4 w}{\partial x^4} + \frac{\rho}{K}\left(I_\rho\frac{\partial^4 w}{\partial t^4} - EI\frac{\partial^4 w}{\partial t^2\,\partial x^2}\right) = 0,$$

which is seen to be a singular perturbation of the Rayleigh equation with ρ/K as the small parameter. Similar, but more complex, results can be obtained in the variable coefficient case and the boundary conditions may be reexpressed in terms of w also.

A model related to that of Timoshenko, but more comprehensive, has recently been developed by Antman [1], [2], and there are many other contributions. In this regard, from our point of view, it is preferable to let the beams "speak for themselves." Figure 4.2 shows the logarithmic power spectrum for excited vibrations of a thin steel beam, measured in the laboratory. The peaks evident in the graph correspond to the natural frequencies of vibration. In this case they conform quite nicely to those predicted from the Euler–Bernoulli theory with very slight corrections from the Rayleigh model proving helpful in the higher ranges.

4.2. Internal Damping Mechanisms.

In traditional applications it is possible to ignore internal damping effects to a large degree. This is the case because, in those applications, interest is largely confined to static deformation or, at a slightly more sophisticated level, to the vibrations produced by periodically imposed loads, such as might be produced, for example, by a train of heavy cars crossing a bridge or a powerful rotating engine mounted on an aircraft wing. Damping effects in these cases can be largely ignored because the standard rule of design is to keep the natural frequencies of vibration of the structure well separated from the frequencies of expected periodic loads.

As aircraft have grown ever more sophisticated and, particularly, as the construction of platforms in space has been contemplated, it has become

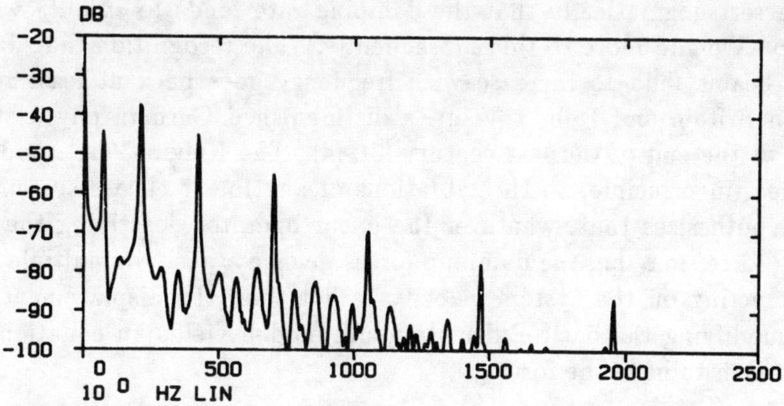

FIG. 4.2

increasingly necessary to take internal energy losses into account. As structures have grown lighter and stronger they have also become more elastic and hence more subject to vibration problems. At the same time, pointing accuracy and platform stability requirements have made the suppression of such vibrations absolutely essential. It has become necessary to augment natural damping by the introduction of active control mechanisms and/or carefully matched passive damping systems. Paradoxically, as more elastic structures with less inherent damping have evolved, and as active control has increasingly been brought to bear, the importance of a thorough understanding of natural damping mechanisms and of a capability to estimate its effects accurately has become more, rather than less, essential. This is true because artificial damping achieved through active control implementation is inevitably limited in bandwidth; at higher frequencies its effectiveness decreases or disappears entirely—indeed, phase lags may result in positive, rather than negative, work being done by the control forces on the higher frequency modes. That the controls used are generally effective is largely thanks to the fact that natural damping mechanisms tend to become stronger with increasing frequency, so that artificial controls can concentrate on suppression of the modes associated with lower frequencies, leaving higher frequency "spillover" effects to be "mopped up" by the strong internal damping operative at the higher end of the spectrum. As internal damping decreases, it is even more important to be able to estimate its precise strength so that the bandwidth requirements for the applied active controls can be appropriately specified.

That damping rates tend to increase with frequency has long been understood [5], [15], [16]. Thus, simplistic models for damping such as

$$\rho\frac{\partial^2 w}{\partial t^2} + 2\gamma\frac{\partial w}{\partial t} + \frac{\partial^2}{\partial x^2}\left(EI\frac{\partial^2 w}{\partial x^2}\right) = 0,$$

which produce uniform damping rates, are generally not adequate if realistic damping properties are to be incorporated in the model. (However, one cannot assert dogmatically that the damping rate tends to *infinity* with the frequency. We add more to this subsequently.) The recognition that damping rates in beams tend to increase with frequency goes back at least to Lord Kelvin in Britain and Robert Voigt, a distinguished German physicist, both working at the end of the last century [3], [4]. The Kelvin–Voigt model may be applied, in principle, to the vibrations of any linear elastic system. The model hypothesizes that, whatever the linear operator describing the elastic restoring forces may be, the damping forces involve a positive multiple of that operator acting on the system velocity rather than the displacement. This theory, modifying the basic Euler–Bernoulli model, yields an equation which may be taken to have the form

$$\rho\frac{\partial^2 w}{\partial t^2} + 2\gamma\rho\frac{\partial^3}{\partial t\,\partial x^2}\left(EI\frac{\partial^2 w}{\partial x^2}\right) + \frac{\partial^2}{\partial x^2}\left(EI\frac{\partial^2 w}{\partial x^2}\right) = 0$$

with appropriately modified boundary conditions. If the fourth-order elasticity operator is denoted by

$$Aw \equiv \frac{1}{\rho}\frac{\partial^2}{\partial x^2}\left(EI\frac{\partial^2 w}{\partial x^2}\right),$$

then the operational form of the equation is

$$\frac{d^2 w}{dt^2} + 2\gamma A\frac{dw}{dt} + Aw = 0. \tag{4.10}$$

If the eigenvalues of the positive selfadjoint operator A are λ_k, so that the natural frequencies of the undamped system are $\omega_k = (\lambda_k)^{1/2}$, then the damped system (4.10) may be seen to have exponential solutions $e^{\sigma_k t}\varphi_k$, where φ_k is the corresponding eigenvector of A and σ_k satisfies the quadratic equation

$$\sigma_k^2 + 2\gamma\omega_k^2\sigma_k + \omega_k^2 = 0.$$

Thus σ_k is given by

$$\sigma_k = -\gamma\omega_k^2 \pm \left(\gamma^2\omega_k^4 - \omega_k^2\right)^{1/2}. \tag{4.11}$$

Assuming γ to be small, we see that the σ_k are complex for some finite number of values of k with the damping rate proportional to the square of the frequency. Critical damping occurs for $\omega = 1/\gamma$; for ω_k larger than this the modes are overdamped with one of the values given by (4.11) going to $-\infty$ and the other tending to the value $-1/2\gamma$. The values σ_k, in fact, lie in the locus shown in Fig. 4.3, consisting of a circle of radius $1/2\gamma$, centered at the point $(-1/2\gamma, 0)$ and the portion of the negative real axis to the left of $(-1/\gamma, 0)$. For very small values of γ, all we would expect to see would be the quadratic dependence of the damping rate on the frequency. Whether the overdamping predicted by the model has ever been observed in the laboratory is unknown to this writer but it seems, on the face of it, to be unlikely.

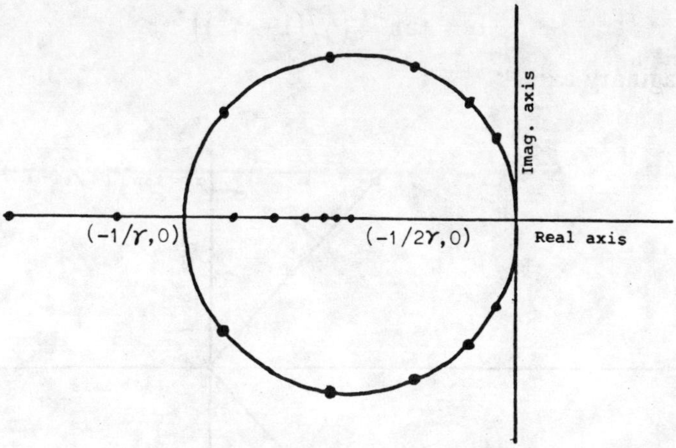

FIG. 4.3

In any case, as we will document in the final section of this chapter, our laboratory experiments, in agreement with experimental data acquired by many other researchers, indicates a predominantly linear relationship between the damping rate and the frequency; a behavior widely known as *structural damping* [13]. We will also see, from the experimental and theoretical point of view, that damping of this sort cannot be considered a "rule of nature"; indeed, there is some reason to believe that it may only apply to thin beams for which the Euler–Bernoulli model is adequate in the conservative case. More on this later.

There is no difficulty in obtaining mathematical models which exhibit this sort of damping/frequency relationship. Representing an arbitrary linear oscillator, without damping, by

$$\frac{d^2w}{dt^2} + Aw = 0,$$

the simplest mathematically viable example of a system exhibiting structural damping behavior, treated extensively by the author and G. Chen in [6], is given by

$$\frac{d^2w}{dt^2} + 2\gamma A^{1/2}\frac{dw}{dt} + Aw = 0 \tag{4.12}$$

where $A^{1/2}$ denotes the positive square root of the positive selfadjoint operator A and $\gamma > 0$. Attempting a solution of the form $e^{\sigma_k t}\varphi_k$ again, we find now that

$$\sigma_k = \left(-\gamma \pm (\gamma^2 - 1)^{1/2}\right)\omega_k$$

so that, for $0 < \gamma < 1$, the σ_k lie on the rays in the left half plane, shown in Fig. 4.4, making an angle

$$\alpha = \tan^{-1}\left(\gamma/(1-\gamma^2)\right)$$

with the imaginary axis.

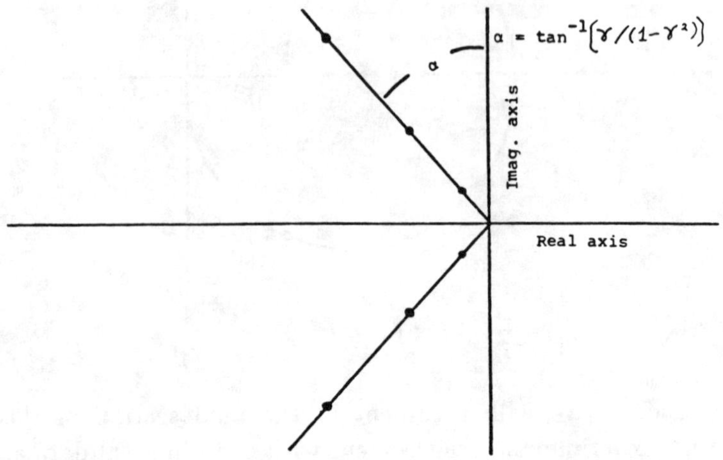

FIG. 4.4

To explain why we cannot embrace the model (4.12) with unrestrained enthusiasm, let us consider the constant coefficient case wherein

$$(4.13) \qquad Aw = \frac{EI}{\rho}\frac{\partial^4 w}{\partial x^4}.$$

Let us denote boundary conditions for this operator generically by

$$(4.14) \qquad \langle \ell, w \rangle = 0,$$

where ℓ is a linear functional defined on the domain of A. We will say that the boundary conditions given for A are *paired* if for each boundary condition (4.14) appearing in the set, the condition

$$(4.15) \qquad \langle \ell, w'' \rangle = 0$$

appears in the set also. Examples include hinged endpoints, for which (taking the endpoint in question to be $x = 0$)

$$w(0,t) = 0, \qquad EI\frac{\partial^2 w}{\partial x^2}(0,t) = 0,$$

and for "guided" endpoints for which (constant coefficient case)

$$\frac{\partial w}{\partial x}(0,t) = 0, \qquad -EI\frac{\partial^3 w}{\partial x^3}(0,t) = 0.$$

For paired boundary conditions the positive square root of the fourth-order operator $\partial^4 w/\partial x^4$ is $-\partial^2 w/\partial x^2$ and, redefining γ slightly, the modified Euler equation in this case takes the form

$$\rho \frac{\partial^2 w}{\partial t^2} - 2\gamma \frac{\partial^3 w}{\partial t\, \partial x^2} + EI \frac{\partial^4 w}{\partial x^4} = 0. \tag{4.16}$$

In this equation the damping term is rather easy to understand from the physical point of view—it is a lateral force acting on the beam which is negatively proportional to the bending rate.

In [21] the set of all natural boundary conditions is shown to be divisible into two classes; with boundary conditions from the first class, which includes the paired boundary conditions we have just discussed, the fourth derivative operator has the negative second derivative as its positive square root. If at either end of the interval $[0, L]$ the fourth-order operator is assigned boundary conditions from the second class, which includes the cases of clamped or free endpoints, corresponding, respectively, to

$$w(0,t) = 0, \qquad \frac{\partial w}{\partial x}(0,t) = 0,$$

$$EI \frac{\partial^2 w}{\partial x^2}(0,t) = 0, \qquad -EI \frac{\partial^3 w}{\partial x^3}(0,t) = 0,$$

for example, the positive square root of the fourth-order operator is not the negative second-order operator and the nature of the damping term in (4.12) is such as not to admit a ready physical interpretation in terms of the properties of standing waves in the beam. If the motion is conceived in terms of travelling waves reflecting from the boundaries of the beam then, since the speed of travelling waves in the Euler–Bernoulli beam varies inversely with the square of the wavelength, "square root" damping can be interpreted, regardless of the boundary conditions in force, as a statement that the damping exponent is proportional to the distance covered per unit time by a travelling wave of the particular wavelength in question. However, this explanation seems rather suspect to the writer because the expression of standing waves in terms of travelling waves is somewhat of a mathematical artifice here. The explanation would become quite appealing, however, if it could be demonstrated that in long thin beams the attenuation of travelling waves is specifically a function of distance travelled rather than other factors. The author is, at the present time, unaware of any research which may have been carried out in this direction.

The guiding philosophy in our modeling work is that a proposed mathematical model for a physical system must:

(i) Be well posed from the mathematical point of view;

(ii) Replicate the physical phenomena observed;

(iii) Involve only equations all of whose terms can be assigned a direct physical meaning in terms of observable system properties or characteristics.

Where the boundary conditions on the beam are not paired, the model (4.12), at least as far as we understand it now, does not meet the third criterion stated and must, consequently, not be accepted as a viable model at the present time.

The apparent necessity of discarding this model under the circumstances just described represents a real loss because the system (4.12) has very attractive mathematical properties, as outlined in [6].

Since the middle term in (4.16) has a ready physical explanation as the rate of change of bending, it is tempting to try to use this equation whatever its disadvantages. If no modifications of the boundary conditions are made, as compared with the Euler–Bernoulli conditions, it is not true in general that the energy is monotone decreasing. Consider, for example, the cantilever case with $L = 1$ and the initial state

$$w(x,0) \equiv 0, \qquad \frac{\partial w}{\partial t}(x,0) = x^2 - \frac{2}{3}x^3 + \frac{1}{6}x^4.$$

(Note that $\partial w/\partial t$ is only required to lie in $L^2[0,L]$ for finite energy solutions, i.e., solutions in the state space $H^2[0,L] \times L^2[0,L]$ appropriate to the corresponding semigroup, and thus need not satisfy the boundary conditions imposed on w at the free endpoint $x = L$.) Then, with $\mathcal{E}$ as in (4.1), we easily compute that

$$\frac{d}{dt}\mathcal{E}\left(w(\cdot,t), \frac{\partial w}{\partial t}(\cdot,t)\right)\bigg|_{t=0} = 2\gamma \int_0^L \frac{\partial w}{\partial t}(x,0)\frac{\partial^3 w}{\partial t\,\partial x^2}(x,0)\,dx = \frac{1}{21}$$

so that the energy is actually increasing during some interval after $t = 0$, rather than decreasing. A further example occurs in [21].

Undaunted by this, it is tempting next to see if some modification of the Euler–Bernoulli boundary conditions will result in solutions of (4.16) having monotone decreasing energy at all times. One can realize this by modifying the zero lateral force condition applying at an endpoint to

$$(4.17) \qquad -2\gamma\frac{\partial^2 w}{\partial t\,\partial x} - EI\frac{\partial^3 w}{\partial x^3} = 0$$

and one then finds, for smooth solutions, that the rate of change of energy is equal to

$$(4.18) \qquad -2\gamma\int_0^L \left(\frac{\partial^2 w}{\partial t\,\partial x}\right)^2 dx \le 0.$$

Unfortunately, our satisfaction is short-lived, for we realize that the beam with free endpoints has purely inertial solutions

$$w(x,t) = w_0 + w_1 t + w_2 x + w_3 t x,$$

and these are not in agreement with the proposed boundary condition (4.17), corresponding to zero lateral force. Nevertheless, the exercise is not quite a total loss. If we work a little harder and change (4.17) to

$$-EI\frac{\partial^3 w}{\partial x^3} + 2\gamma\frac{\partial^2 w}{\partial t\,\partial x} - \frac{2\gamma}{L}\left(\frac{\partial w}{\partial t}(L,t) - \frac{\partial w}{\partial t}(0,t)\right) = 0, \tag{4.19}$$

we then find that the purely inertial solutions are admitted and, for smooth solutions, the energy decay rate becomes

$$\frac{d}{dt}\mathcal{E}\left(w(\cdot,t), \frac{\partial w}{\partial t}(\cdot,t)\right)$$
$$= -2\gamma\int_0^L\left(\frac{\partial^2 w}{\partial t\,\partial x}(x,t)\right)^2 dx + \frac{2\gamma}{L}\left(\frac{\partial w}{\partial t}(L,t) - \frac{\partial w}{\partial t}(0,t)\right)^2;$$

the energy is seen to be nonincreasing for all such solutions while remaining constant for the purely inertial motions. Condition (4.19) seems somewhat unnatural since it is nonlocal. In fact, we do not propose (4.19) as a completely serious solution to our problem, but this small success leads in a natural way to the model, which we hope the reader will find more believable, to be discussed in the next section.

4.3. Form of the Mathematical Model.

As we have noted in the previous section, the commonly used Euler–Bernoulli beam equation, which will serve as the starting point for our discussion here, is based on conservation in time of the energy expression (4.1), i.e.,

$$\mathcal{E}\left(w, \frac{\partial w}{\partial t}\right) = \frac{1}{2}\int_0^L\left[\rho\left(\frac{\partial w}{\partial t}\right)^2 + EI\left(\frac{\partial^2 w}{\partial x^2}\right)^2\right] dx,$$

wherein $w(x,t)$ is abbreviated to w, and $\rho(x)$ and $EI(x)$ to ρ and EI, respectively. We will assume here that ρ and EI are uniformly positive on $0 \leq x \leq L$, that ρ is continuous, at least, and that EI is at least piecewise twice continuously differentiable on $[0, L]$.

If we assume for the moment that the function $w(x,t)$ describing the evolution of the beam displacement is smooth, an easy calculation shows that

$$\frac{d}{dt}\mathcal{E}\left(w(\cdot,t), \frac{\partial w}{\partial t}(\cdot,t)\right) = \int_0^L\left[\rho\frac{\partial w}{\partial t}\,\frac{\partial^2 w}{\partial t^2} + EI\frac{\partial^2 w}{\partial x^2}\,\frac{\partial^3 w}{\partial t\,\partial x^2}\right] dx \tag{4.20}$$

and then, integrating the second term by parts,

(4.21)
$$\frac{d}{dt}\mathcal{E}\left(w(\cdot,t), \frac{\partial w}{\partial t}(\cdot,t)\right)$$
$$= \int_0^L\left[\rho\frac{\partial w}{\partial t}\,\frac{\partial^2 w}{\partial t^2} - \underline{\frac{\partial}{\partial x}\left(EI\frac{\partial^2 w}{\partial x^2}\right)\frac{\partial^2 w}{\partial t\,\partial x}}\right] dx + EI\,\frac{\partial^2 w}{\partial x^2}\,\frac{\partial^2 w}{\partial t\,\partial x}\bigg|_0^L.$$

The presence of the angular velocity expression $\partial^2 w/\partial t\, \partial x$ in the underlined term indicates that its coefficient, $-(\partial/\partial x)(EI\partial^2 w/\partial x^2)$, should be interpreted as a restoring *torque*, arising due to spatially variable bending of the beam. Realizing that this coefficient represents a torque aids us in interpretation of the damping term which we now introduce into the system via the definition

$$(4.22) \qquad \tau_h(x,t) = 2\int_0^L h(x,\xi)\left[\frac{\partial^2 w}{\partial t\,\partial x}(x,t) - \frac{\partial^2 w}{\partial t\,\partial x}(\xi,t)\right] d\xi.$$

We think of τ_h as a torque acting on the beam at the point x due to the differential rotation, as compared with the rotation at x, of the beam at points ξ "near" x. In many cases the support of the "interaction kernel" $h(x,\xi)$ would be restricted to a thin strip in R^2, centered on the line $x = \xi$. Application of Newton's second law dictates the symmetry condition

$$(4.23) \qquad h(\xi,x) = h(x,\xi).$$

In constant coefficient applications it is convenient to replace $h(x,\xi)$ by $\gamma h(x-\xi)$, where $\gamma > 0$ is used to parametrize the strength of the damping effect and $h(\eta)$ satisfies the normalization condition

$$(4.24) \qquad \int_{-\delta}^{\delta} h(\eta)\, d\eta = 1$$

and the evenness condition $h(\eta) = h(-\eta)$.

The source of the damping torque in differential rotation is best illustrated for the case of beams composed of composite materials, such as fiberglass, boron and graphite composites, and wood. We may imagine that long fibers, whose modulus of elasticity, per unit cross-sectional area, is greater than that of the beam as a whole, pass through the beam, held in place by a matrix material of some sort. As the beam undergoes deformation of various sorts, beam elements at x and ξ may rotate at different rates, reflected by different values of $(\partial^2 w/\partial t\, \partial x)(x,t)$, $(\partial^2 w/\partial t\, \partial x)(\xi,t)$. If we think of the fibers themselves as having nearly constant length, differential rotation must result in movement of the fibers relative to the matrix, with accompanying frictional force, or deformation (largely inelastic) of the matrix material. The result is a torque of the type which we have just described, since the net motion of the fibers relative to the matrix, within the individual beam elements, will be different on one side of the elastic axis than on the other side when such differential rotation does take place (see Fig. 4.5).

We add the term

$$\int_0^L \tau_h(x,t)\frac{\partial^2 w}{\partial t\,\partial x}(x,t)\, dx$$

to both sides of (4.21), so that

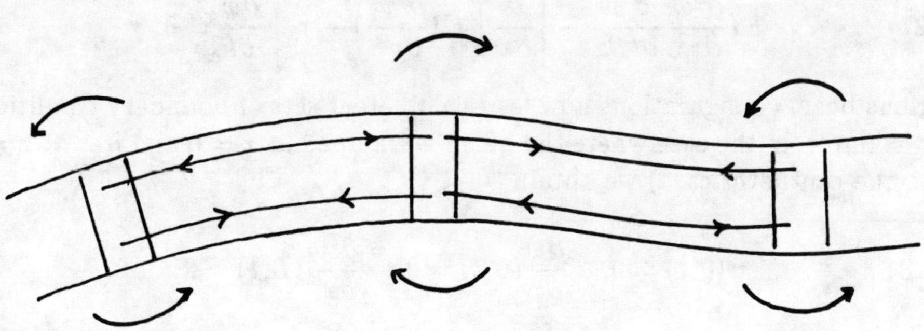

FIG. 4.5

(4.25)

$$
\begin{aligned}
&\frac{d}{dt}\mathcal{E}\left(w(\cdot,t),\frac{\partial w}{\partial t}(\cdot,t)\right)+\int_0^L \tau_h\frac{\partial^2 w}{\partial t\,\partial x}dx\\
&=\int_0^L\left\{\rho\frac{\partial w}{\partial t}\frac{\partial^2 w}{\partial t^2}+\left[\tau_h-\frac{\partial}{\partial x}\left(EI\frac{\partial^2 w}{\partial x^2}\right)\right]\frac{\partial^2 w}{\partial t\,\partial x}\right\}dx\\
&+EI\frac{\partial^2 w}{\partial x^2}\frac{\partial^2 w}{\partial t\,\partial x}\bigg|_0^L=\text{(with a further integration by parts)}\\
&=\int_0^L\frac{\partial w}{\partial t}\left\{\rho\frac{\partial^2 w}{\partial t^2}+\frac{\partial}{\partial x}\left[\frac{\partial}{\partial x}\left(EI\frac{\partial^2 w}{\partial x^2}\right)-\tau_h\right]\right\}dx\\
&+\left\{EI\frac{\partial^2 w}{\partial x^2}\frac{\partial^2 w}{\partial t\,\partial x}-\left[\frac{\partial}{\partial x}\left(EI\frac{\partial^2 w}{\partial x^2}\right)-\tau_h\right]\frac{\partial w}{\partial t}\right\}\bigg|_0^L.
\end{aligned}
$$

Equating the separate parts of (4.25) to zero (a procedure that can be justified by the principal of virtual work) yields the integro–partial differential equation

(4.26)

$$
\begin{aligned}
&\rho\frac{\partial^2 w}{\partial t^2}+\frac{\partial}{\partial x}\left[\frac{\partial}{\partial x}\left(EI\frac{\partial^2 w}{\partial x^2}\right)-\tau_h\right]\\
&=\rho\frac{\partial^2 w}{\partial t^2}-2\frac{\partial}{\partial x}\int_0^L h(x,\xi)\left[\frac{\partial^2 w}{\partial t\,\partial x}(x,t)-\frac{\partial^2 w}{\partial t\,\partial x}(\xi,t)\right]d\xi+\frac{\partial^2}{\partial x^2}\left(EI\frac{\partial^2 w}{\partial x^2}\right)\\
&=0
\end{aligned}
$$

and the requirement that, at $x = 0$ and $x = L$,

$$EI\frac{\partial^2 w}{\partial x^2}\frac{\partial^2 w}{\partial t\,\partial x} - \left[\frac{\partial}{\partial x}\left(EI\frac{\partial^2 w}{\partial x^2}\right) - \tau_h\right]\frac{\partial w}{\partial t} = 0. \tag{4.27}$$

Various beam configurations now lead to different sets of boundary conditions. For example, in the case where the beam is clamped at $x = 0$ and free at $x = L$ (i.e., the cantilever case) we obtain

$$w(0,t) = 0, \quad \frac{\partial w}{\partial x}(0,t) = 0, \quad \frac{\partial^2 w}{\partial x^2}(L,t) = 0, \tag{4.28}$$

(4.29)

$$-\frac{\partial}{\partial x}\left(EI(x)\frac{\partial^2 w}{\partial x^2}\right)\bigg|_{x=L} + 2\int_0^L h(L,\xi)\left[\frac{\partial^2 w}{\partial t\,\partial x}(L,t) - \frac{\partial^2 w}{\partial t\,\partial x}(\xi,t)\right]d\xi = 0.$$

The reader will be able to generate boundary conditions corresponding to other configurations with equal ease.

Equation (4.25) now becomes

$$\begin{aligned}
&\frac{d}{dt}\mathcal{E}\left(w(\cdot,t), \frac{\partial w}{\partial t}(\cdot,t)\right)\\
&\quad = -2\int_0^L\int_0^L h(x,\xi)\left[\frac{\partial^2 w}{\partial t\,\partial x}(\xi,t) - \frac{\partial^2 w}{\partial t\,\partial x}(x,t)\right]d\xi\,\frac{\partial^2 w}{\partial t\,\partial x}(x,t)\,dx\\
&\quad = \text{(since the roles of } x \text{ and } \xi \text{ are symmetric and } h(x,\xi) = h(\xi,x))\\
&\quad = -\int_0^L\int_0^L h(\xi,x)\left[\frac{\partial^2 w}{\partial t\,\partial x}(x,t) - \frac{\partial^2 w}{\partial t\,\partial x}(\xi,t)\right]dx\,\frac{\partial^2 w}{\partial t\,\partial x}(\xi,t)\,d\xi\\
&\qquad -\int_0^L\int_0^L h(x,\xi)\left[\frac{\partial^2 w}{\partial t\,\partial x}(\xi,t) - \frac{\partial^2 w}{\partial t\,\partial x}(x,t)\right]d\xi\,\frac{\partial^2 w}{\partial t\,\partial x}(x,t)dx \qquad (4.30)\\
&\quad = -\int_0^L\int_0^L h(x,\xi)\left[\frac{\partial^2 w}{\partial t\,\partial x}(x,t) - \frac{\partial^2 w}{\partial t\,\partial x}(\xi,t)\right]\frac{\partial^2 w}{\partial t\,\partial x}(\xi,t)\,d\xi\,dx\\
&\qquad + \int_0^L\int_0^L h(x,\xi)\left[\frac{\partial^2 w}{\partial t\,\partial x}(x,t) - \frac{\partial^2 w}{\partial t\,\partial x}(\xi,t)\right]\frac{\partial^2 w}{\partial t\,\partial x}(x,t)d\xi\,dx\\
&\quad = -\int_0^L\int_0^L h(x,\xi)\left[\frac{\partial^2 w}{\partial t\,\partial x}(x,t) - \frac{\partial^2 w}{\partial t\,\partial x}(\xi,t)\right]^2 d\xi\,dx \le 0.
\end{aligned}$$

Moreover, assuming that $h(x,\xi)$ is not identically zero, the inequality is strict except in precisely the cases where $\partial^2 w/\partial t\,\partial x$ is constant, i.e., the inertial motions. Thus energy is strictly decreasing whenever the bending rate is not identically zero and is conserved when the bending rate, $\partial^3 w/\partial t\,\partial x^2$, vanishes identically.

In general the form of $h(x,\xi)$ will depend on the elasticity of the fibers of the material, as discussed earlier, the distribution of their lengths, the nature of the interface between them and the matrix material, etc. The more elastic the fibers and the shorter their average length, the more the "mass" of $h(x,\xi)$ will be concentrated near the line $\xi = x$.

For perfectly inelastic fibers, all having length L, the same length as the beam itself, $h(x,\xi)$ is a constant; call its value γ/L. Then

$$\int_0^L h(x,\xi)\frac{\partial^2 w}{\partial t\,\partial x}(\xi,t)d\xi \equiv \frac{\gamma}{L}\left(\frac{\partial w}{\partial t}(L,t) - \frac{\partial w}{\partial t}(0,t)\right),$$

$$\int_0^L h(x,\xi)\frac{\partial^2 w}{\partial t\,\partial x}(x,t)d\xi \equiv \gamma\frac{\partial^2 w}{\partial t\,\partial x}(x,t),$$

so that (4.26) becomes

$$\rho\frac{\partial^2 w}{\partial t^2} - 2\gamma\frac{\partial^3 w}{\partial t\,\partial x^2} + \frac{\partial^2}{\partial x^2}\left(EI\frac{\partial^2 w}{\partial x^2}\right) = 0$$

and the fourth boundary condition of the cantilever case in (4.29), for example, becomes

$$-\frac{\partial}{\partial x}\left(EI(x)\frac{\partial^2 w}{\partial x^2}\right)\bigg|_{x=L} + 2\gamma\frac{\partial^2 w}{\partial t\,\partial x}(L,t) - \frac{2\gamma}{L}\frac{\partial w}{\partial t}(L,t) = 0,$$

in exact agreement with the model obtained "ad hoc" at the end of the previous section.

In the case of metallic beams, the "fiber" explanation used above is not persuasively valid. It may be possible to think of elongated crystal structures in the material playing much the same role as the fibers above, but the analogy may well be far-fetched. The evidence for frequency-proportional damping is not quite as compelling at this writing for metallic beams as it is for composite and wooden beams, as we will see in our discussion of experimental results in §4.9. It may turn out, for metallic beams, that the thermodynamic damping mechanisms discussed, e.g., by Zener in [23] are more to the point. We intend to examine these in detail elsewhere. Suffice it to say that the thermodynamic model leads to damping rates proportional to the square of the frequency at the low end of the spectrum, as in the Kelvin–Voigt model, with asymptotically constant damping rates being approached at the high frequency end. The thermodynamic model leads to a partial differential/delay equation. More general models of this type have been discussed by Hannsgen and Wheeler in [11] and by Hannsgen, Renardy and Wheeler in [12].

4.4. Additional Properties of the Model.

Application to metallic beams, if valid at all, would necessarily involve the supposition that the fiber/crystal structures, in terms of which the damping action has been explained, are rather short. It is therefore of some interest to investigate the limiting form of the equations, as the support of $h(x,\xi)$ is restricted to small neighborhoods of the line $x = \xi$.

To avoid complications with the boundary conditions here, let us consider an infinite beam, so that 0 is replaced by $-\infty$ and L by ∞. Then the energy loss term in (4.30) is

$$\int_{-\infty}^{\infty}\int_{-\infty}^{\infty} h(x,\xi)\left[\frac{\partial^2 w}{\partial t\,\partial x}(x,t) - \frac{\partial^2 w}{\partial t\,\partial x}(\xi,t)\right]^2 d\xi\,dx. \tag{4.31}$$

Using the change of variables

$$y = \frac{1}{\sqrt{2}}(x+\xi), \qquad \eta = \frac{1}{\sqrt{2}}(x-\xi),$$

the integral (4.31) may be rewritten as

$$\int_{-\infty}^{\infty}\int_{-\infty}^{\infty} h\left(\frac{1}{\sqrt{2}}(y+\eta), \frac{1}{\sqrt{2}}(y-\eta)\right)$$
$$\cdot\left[\frac{\partial^2 w}{\partial t\,\partial x}\left(\frac{1}{\sqrt{2}}(y+\eta),t\right) - \frac{\partial^2 w}{\partial t\,\partial x}\left(\frac{1}{\sqrt{2}}(y-\eta),t\right)\right]^2 d\eta\,dy.$$

Now, assuming w sufficiently smooth,

$$\frac{\partial^2 w}{\partial t\,\partial x}\left(\frac{1}{\sqrt{2}}(y\pm\eta),t\right) = \frac{\partial^2 w}{\partial t\,\partial x}\left(\frac{y}{\sqrt{2}},t\right)$$
$$\pm\frac{\eta}{\sqrt{2}}\int_0^1 \frac{\partial^2 w}{\partial t\,\partial x^2}\left(\frac{1}{\sqrt{2}}(y\pm\sigma\eta),t\right) d\sigma$$

and the integral is

$$2\int_{-\infty}^{\infty}\int_{-\infty}^{\infty} h\left(\frac{1}{\sqrt{2}}(y+\eta), \frac{1}{\sqrt{2}}(y-\eta)\right)\eta^2$$
$$\cdot\left[\int_0^1 \frac{\partial^3 w}{\partial t\,\partial x^2}\left(\frac{1}{\sqrt{2}}(y\pm\sigma\eta),t\right) d\sigma\right]^2 d\eta\,dy. \tag{4.32}$$

Consider a family of functions

$$h(x,\xi) = h_\gamma(x,\xi), \qquad \gamma > 0,$$

each having the properties assumed earlier for $h(x,\xi)$ and such that

$$\lim_{\gamma\to 0}\int_{-\gamma}^{\gamma} h_\gamma\left(\frac{1}{\sqrt{2}}(y+\eta),\frac{1}{\sqrt{2}}(y-\eta)\right)\eta^2\,d\eta \equiv \frac{1}{\sqrt{2}}H\left(\frac{y}{\sqrt{2}}\right)$$

while

$$\left(\int_{-\infty}^{\gamma}+\int_{\gamma}^{\infty}\right)h_\gamma\left(\frac{1}{\sqrt{2}}(y+\eta),\frac{1}{\sqrt{2}}(y-\eta)\right)\eta^2\,d\eta \le J(\gamma)G(y),$$

where $H(y)$ and $G(y)$ are positive integrable functions of y and

$$\lim_{\gamma\to 0} J(\gamma) = 0.$$

Assuming that $\partial^3 w/\partial t\,\partial x^2$ is such that

$$\left[\int_0^1 \frac{\partial^3 w}{\partial t\,\partial x^2}\left(\frac{1}{\sqrt{2}}(y\pm\sigma\eta),t\right)d\sigma\right]^2 \le B$$

uniformly for all y and η while

$$\left|\left(\frac{\partial^3 w}{\partial t\,\partial x^2}\left(\frac{y}{\sqrt{2}},t\right)\right)^2 - \left[\int_0^1 \frac{\partial^3 w}{\partial t\,\partial x^2}\left(\frac{1}{\sqrt{2}}(y\pm\sigma\eta),t\right)d\sigma\right]^2\right| \le \delta(\gamma)$$

uniformly for all y in $(-\infty,\infty)$ and for η in $[-\gamma,\gamma]$, where

$$\lim_{\gamma\to 0}\delta(\gamma) = 0,$$

the limiting value, as $\gamma\to 0$, of the integral (4.32) is seen to be

$$\begin{aligned}&\sqrt{2}\int_{-\infty}^{\infty} H\left(\frac{1}{\sqrt{2}}\right)\left[\frac{\partial^3 w}{\partial t\,\partial x^2}\left(\frac{y}{\sqrt{2}},t\right)\right]^2 dy\\ &= 2\int_{-\infty}^{\infty} H(x)\left[\frac{\partial^3 w}{\partial t\,\partial x^2}(x,t)\right]^2 dx.\end{aligned} \tag{4.33}$$

Returning to equation (4.20) with $0, L$ replaced by $-\infty$, ∞, respectively, we see that, if the term (4.33) is added to both sides of that equation, there results

$$\begin{aligned}&\frac{d}{dt}\mathcal{E}\left(w(\cdot,t),\frac{\partial w}{\partial t}(\cdot,t)\right)+2\int_{-\infty}^{\infty}H(x)\left(\frac{\partial^3 w}{\partial t\,\partial x^2}(x,t)\right)^2 dx\\ &= \int_{-\infty}^{\infty}\left[\rho\frac{\partial w}{\partial t}\,\frac{\partial^2 w}{\partial t^2}+\left(2H\frac{\partial^3 w}{\partial t\,\partial x^2}+EI\frac{\partial^2 w}{\partial x^2}\right)\frac{\partial^3 w}{\partial t\,\partial x^2}\right]dx\end{aligned}$$

$$= \text{(after integrating the last product twice by parts with respect to } x\text{)}$$

$$= \int_{-\infty}^{\infty} \frac{\partial}{\partial t}\left[\rho \frac{\partial^2 w}{\partial t^2} + 2\frac{\partial^2}{\partial x^2}\left(H\frac{\partial^3 w}{\partial t\,\partial x^2}\right) + \frac{\partial}{\partial x^2}\left(EI\frac{\partial^2 w}{\partial x^2}\right)\right] dx.$$

Here we have, of course, assumed that w, $\partial w/\partial t$, and their x-derivatives tend to zero at an appropriate rate as $|x| \to \infty$.) Now the condition that the integrand should vanish identically corresponds to the Kelvin–Voigt partial differential equation

$$\rho \frac{\partial^2 w}{\partial t^2} + 2\frac{\partial^2}{\partial x^2}\left(H\frac{\partial^3 w}{\partial t\,\partial x^2}\right) + \frac{\partial^2}{\partial x^2}\left(EI\frac{\partial^2 w}{\partial x^2}\right) = 0. \tag{4.34}$$

Thus our model is consistent with the Kelvin–Voigt model, under the stated conditions, as the "mass" of $h(x,\xi)$ is progressively more concentrated near $x = \xi$. For fixed γ, equivalently fixed $h(x,\xi)$, this means that our model approximates the Kelvin–Voigt model for vibrations of large wavelength, i.e., low frequency. This will be made more precise in the following material on the spectral properties of the constant coefficient equation on $(-\infty,\infty)$.

Further work of a more precise character will be required in order to obtain a rigorous argument showing that solutions of (4.26) approach solutions of (4.34) as $\gamma \to 0$ and to account for the same phenomena in finite beams with their attendant boundary conditions. There are some preliminary developments in this direction, but the work must appear elsewhere.

We pass now to a discussion of the spectral properties of the constant coefficient equation on $(-\infty,\infty)$. Taking our earlier remarks into account, the constant coefficient version of our equation, with $\eta = x - \xi$, is

(4.35)

$$\rho \frac{\partial^2 w}{\partial t^2} - 2\gamma \frac{\partial}{\partial x}\int_{-\infty}^{\infty} h(\eta)\left[\frac{\partial^2 w}{\partial t\,\partial x}(x,t) - \frac{\partial^2 w}{\partial t\,\partial x}(x-\eta,t)\right] d\eta + EI\frac{\partial^4 w}{\partial x^4} = 0.$$

The function h has the properties described in (4.23), (4.24) and ρ, EI are positive constants. Equation (4.35) applies for an infinite beam or, in much the same form, for the periodic case wherein x is identified with $x + L$ for some finite $L > 0$. Our purpose in this paragraph is to study spectral properties of this equation under the assumption that $h \in L^1(-\infty,\infty)$ with $\int_{-\infty}^{\infty} h(\eta)d\eta = 1$. Accordingly, our equation becomes

$$\rho \frac{\partial^2 w}{\partial t^2} - 2\gamma \frac{\partial^3 w}{\partial t\,\partial x^2} + \int_{-\infty}^{\infty} h(\eta)\frac{\partial^3 w}{\partial t\,\partial x^2}(x-\eta,t)d\eta + EI\frac{\partial^4 w}{\partial x^4} = 0, \tag{4.36}$$

the form which we will use for our analysis. The periodic case may be developed along quite similar lines.

We begin by looking for solutions of (4.36) in the form

$$w(x,t) = e^{\sigma t}\varphi(x)$$

leading to the steady state equation

$$\rho\sigma^2\varphi(x) - 2\gamma\sigma\varphi''(x) + \sigma\int_{-\infty}^{\infty} h(\eta)\varphi''(x-\eta)d\eta + EI\varphi^{(iv)}(x) = 0. \tag{4.37}$$

Let

$$\psi(\tau) = \int_{-\infty}^{\infty} e^{-i\tau x}\varphi(x)dx, \qquad -\infty < \tau < \infty, \tag{4.38}$$

be the Fourier transform of φ, convergent in an appropriate sense. Applying the Fourier operator to the equation, we have

$$\left(\rho\sigma^2 + 2\gamma\tau^2(1-\hat{h}(\tau))\sigma + EI\tau^4\right)\psi(\tau) = 0. \tag{4.39}$$

The coefficient of $\psi(\tau)$ must vanish on the support of the function (or distribution) $\psi(\tau)$, leading to the following relationship between σ and τ:

$$\sigma = \sigma(\tau) = \frac{\tau^2}{\rho}\left[-g(\tau) \pm \sqrt{g(\tau)^2 - \rho EI}\right], \tag{4.40}$$

with

$$g(\tau) = \gamma(1-\hat{h}(\tau)). \tag{4.41}$$

Here, as implied earlier, $\hat{h}(\tau)$ is the Fourier transform of the interaction kernel $h(\eta)$. From the integrability of $h(\eta)$ we know that $\hat{h}(\tau)$ is continuous and

$$\lim_{|\tau|\to\infty} \hat{h}(\tau) = 0. \tag{4.42}$$

For a relatively low level of damping we may assume $\gamma^2 < \rho EI$. Then, from (4.40)–(4.42), it is clear that for large values of τ

$$\sigma \simeq \frac{\tau^2}{\rho}\left[-\gamma(1-\hat{h}(\tau)) \pm i\sqrt{\rho EI-\gamma^2}\right.$$
$$\left.\cdot\left(1+\frac{\gamma^2\hat{h}(\tau)-\gamma^2(\hat{h}(\tau)^2/2+\cdots}{\rho EI-\gamma^2}\right)\right]$$
$$\simeq \frac{\tau^2}{\rho}\left[-\gamma \pm i\sqrt{\rho EI-\gamma^2}\right]\left[1\mp\frac{1\gamma\hat{h}(\tau)}{\sqrt{\rho EI-\gamma^2}}\right]$$

to first order in $\hat{h}(\tau)$ as $|\tau|\to\infty$. Asymptotically, as $|\tau|\to\infty$, σ lies along the rays

$$R_{+,-}=\left\{z|z=r\left(-\gamma\pm i\sqrt{\rho EI-\gamma^2}\right),\quad r>0\right\}$$

in the left half complex plane.

From (4.24) it is clear that

$$\lim_{|\tau|\to 0}\hat{h}(\tau)=1$$

so that

$$\sigma\simeq\frac{\tau^2}{\rho}\left[\pm i\sqrt{\rho EI-\gamma^2}+\mathcal{O}(1)\right],\qquad |\tau|\to 0,$$

showing that the spectral curve is tangent to the imaginary axis at the origin. In fact, since h is even, two integrations by parts show that

$$\hat{h}(\tau)=\int_{-\infty}^{\infty}e^{-i\tau\eta}h(\eta)d\eta=1-\tau^2\int_{-\infty}^{\infty}e^{-i\tau\eta}\Theta(\eta)d\eta,$$

where

$$\Theta''(\eta)=h(\eta),\qquad \lim_{|\eta|\to\infty}\Theta(\eta)=0,\qquad \lim_{|\eta|\to\infty}\Theta'(\eta)=0.$$

Thus, as $|\tau|\to 0$,

$$\sigma\simeq\frac{\tau^2}{\rho}\left[-\gamma\theta\tau^2\pm i\sqrt{\rho EI-\gamma^2}\right],$$

$$\theta=\int_{-\infty}^{\infty}\Theta(\eta)d\eta=\frac{1}{2}\int_{-\infty}^{\infty}\eta^2h(\eta)d\eta.$$

Thus, at longer wavelengths, the spectrum tends to that of the corresponding Kelvin–Voigt equation, as we should expect from our earlier work in this section, since taking long wavelengths for fixed h may be seen to be equivalent to allowing the support of $h(\eta)$ to contract to $\{0\}$, keeping wavelength fixed. A typical spectral curve for the beam of infinite length with the proposed damping mechanism is shown in Fig. 4.6.

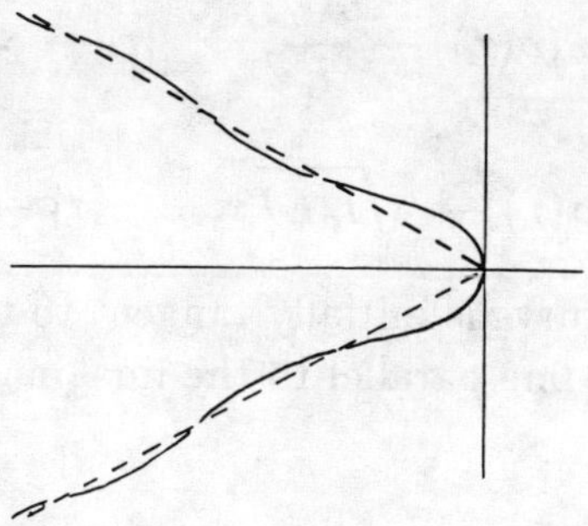

FIG. 4.6

For beams of large finite length L the spectrum will consist of a sequence of discrete points closely spaced along a qualitatively similar curve in the complex plane. The term "large finite length" is, of course, intended to be understood relative to the support, or concentration, of the interaction kernel $h(\eta)$ in our model.

From (4.39) we have

$$\hat{h}(\tau) = 1 + \frac{EI\tau^2}{2\gamma\sigma(\tau)} + \frac{\rho\sigma(\tau)}{2\gamma\tau^2}.$$

Thus, in principle, $\hat{h}(\tau)$, and hence $h(\eta)$, can be reconstructed from the knowledge of $\sigma(\tau)$. This remains true in an approximate sense for finite beams of large length and may prove useful in identifying actual dissipation mechanisms in the laboratory.

It is interesting to note what happens when the basic, conservative, system (before damping is introduced), i.e., the Euler–Bernoulli model in the above analysis, is replaced by the Rayleigh model (4.5). Equation (4.39) is then replaced by

$$\left(\left(\rho + I_\rho\tau^2\right)\sigma^2 + 2\gamma\tau^2\left(1 - \hat{h}(\tau)\right)\sigma + EI\tau^4\right)\psi(\tau) = 0$$

which leads to (cf. (4.40))

$$\begin{aligned}\sigma &= \frac{\tau^2}{\rho + I_\rho\tau^2}\left[-g(\tau) \pm \sqrt{g(\tau)^2 - (\rho + I_\rho\tau^2)\, EI}\right]\\ &\simeq \frac{\tau^2}{\rho + I_\rho\tau^2}\Bigg[-\gamma \pm i\sqrt{(\rho + I_\rho\tau^2)\, EI - \gamma^2}\\ &\qquad \underline{\mp \frac{i\gamma\hat{h}(\tau)}{\sqrt{(\rho + I_\rho\tau^2)\, EI - \gamma^2}}\left(-\gamma \pm i\sqrt{(\rho + I_\rho\tau^2)\, EI - \gamma^2}\right)\Bigg].}\end{aligned}$$

The underlined terms tend to zero as $|\tau| \to \infty$, so that

$$\operatorname{Re}(\sigma(\tau)) \to \frac{-\gamma}{I_\rho}, \qquad |\tau| \to \infty,$$

$$\operatorname{Im}(\sigma(\tau)) \to \sqrt{I_\rho EI}\,\tau, \qquad |\tau| \to \infty.$$

In this case the spectral curve is initially tangent to the imaginary axis, for $|\tau|$ near zero, and tends to a line parallel to the imaginary axis, γ/I_ρ units to the left, as $|\tau| \to \infty$.

4.5. State Space, Boundary Conditions, Semigroup Formulation.

From the form (4.1) of the energy expression for the Euler–Bernoulli beam we are led to define the state space as a subspace, or in some cases a quotient space, of $H^2[0,L] \times L^2[0,L]$. The inner product should be consistent with the bilinear form

$$\langle (w,v),(\hat{w},\hat{v})\rangle = \int_0^L \left[\rho(x)v(x)\bar{\hat{v}}(x) + EI(x)w''(x)\overline{\hat{w}''(x)}\right] dx$$

which, in turn, is related to the energy-motivated seminorm

$$\|(w,v)\|^2 = \int_0^L \left[\rho(x)|v(x)|^2 + EI(x)|w''(x)|^2\right] dx. \tag{4.43}$$

To do this properly it is necessary to say something about the boundary conditions, not all of which are created equal. Let B be the set of distribution pairs, in Schwartz's sense, such that the boundary conditions on the unforced beam are expressible in the form

$$\langle \beta_1, w\rangle + \langle \beta_2, v\rangle = 0, \qquad (\beta_1,\beta_2) \in B. \tag{4.44}$$

Some of these boundary conditions represent kinematic constraints while others are really dynamical equations. The former constitute a subset $C \subset D$ of pairs $(\beta,0)$, where β is a continuous linear functional on $H^2[0,L]$, which prescribe certain constraints on the beam displacement without any reference to applied forces, such as, e.g.,

$$w(0,t) = 0, \qquad \frac{\partial w}{\partial x}(0,t) = 0, \text{ etc.}$$

Let us denote the subspace of $H^2[0,L]$ consisting of w such that $\langle \beta, w\rangle = 0$, $(\beta,0) \in C$, by $H^2_C[0,L]$. The remaining boundary conditions, corresponding to distribution pairs in $B - C$, are really dynamical equations which have to do with certain forces applied to the system or, in the free case, state that certain forces are zero. Thus, in the cantilever case

$$EI\frac{\partial^2 w}{\partial x^2}(L,t) = 0, \qquad -\frac{\partial}{\partial x}\left(EI\frac{\partial^2 w}{\partial x^2}\right)(L,t) = 0$$

are statements that the applied torque and lateral force, respectively, at $x = L$, are zero. In the forced case the right-hand sides would be replaced by a moment μ and a lateral force φ. Although the elements of $B - C$ in this cantilever case still have the form $(\beta, 0)$, it is clear that β is not a continuous linear functional on $H^2[0, L]$.

Let N_C be the subspace of $H_C^2[0, L] \times L^2[0, L]$ on which the seminorm (5.1) vanishes. In many cases $N_C = \{0\}$ but not, e.g., in the case of the beam with two free endpoints, or in the case where one endpoint is free and the other hinged. Extended in the obvious way to

$$\mathcal{H} \equiv \left(H_C^2[0, L] \times L^2[0, L]\right) / N_C$$

the seminorm (5.1) becomes the norm for our state space, $\mathcal{H}$. We will continue to refer to state vectors as displacement and velocity pairs (w, v); the equivalence class will be understood in those cases where $N_C \neq \{0\}$.

The state space now defined, we proceed to pose our dynamical equations in that context. With $z = (w, v)$, consider the unbounded operator on $\mathcal{H}$ defined by

$$\mathbb{A} z = \begin{pmatrix} 0 & I \\ -A & -G \end{pmatrix} \begin{pmatrix} w \\ v \end{pmatrix},$$

where the operational definitions of A and B are

$$(Aw)(x) = (EI(x) w''(x))'', \qquad (Gv)(x) = -2 \frac{\partial}{\partial x} \int_0^L h(x, \xi)[v'(x) - v'(\xi)]\, d\xi.$$

The domain of $\mathbb{A}$ consists of those pairs $(w, v) \in H^4[0, L] \times H^2[0, L]$ such that (4.44) is true for all $(\beta_1, \beta_2) \in B$. Passing to equivalence classes where necessary, this is a dense subspace $\mathcal{D}(\mathbb{A}) \subset \mathcal{H}$ and $\mathbb{A}$, so defined, is a closed operator. Our model equation (4.26) together with the boundary conditions arising from (4.27), and discussed earlier, is then equivalent to

$$\dot{z} = \mathbb{A} z. \tag{4.45}$$

The computation (4.30) can then be used, with a few refinements, to show that $\mathbb{A}$ is dissipative, i.e.,

$$\langle z, \mathbb{A} z \rangle \leq 0, \qquad z \in \mathcal{D}(\mathbb{A}).$$

Using this, one sees (cf. [14], [18]) that $\mathbb{A}$ generates a strongly continuous semigroup, $Z(t)$, of bounded operators on $\mathcal{H}$.

It turns out that the semigroup $Z(t)$ is, in fact, a holomorphic semigroup, as one might expect from the location of the spectrum in the constant coefficient, infinite length case discussed earlier. A general idea of the proof of holomorphicity proceeds in the following way. First, one studies the operator

$$\mathbb{A}_0 = \begin{pmatrix} 0 & I \\ -A & -G_0 \end{pmatrix},$$

where

$$(G_0 v)(x) = -2\frac{\partial}{\partial x}\left(H(x)v'(x)\right), \qquad H(x) = \int_0^L h(x,\xi)\,d\xi,$$

the domain of $\mathbb{A}_0$ being defined in terms of a suitably adjusted set of boundary functionals, B_0. The boundary conditions corresponding to the subset $C \subset B$ remain unchanged as a subset of B_0, so the state space $\mathcal{H}$ remains unchanged. The problem of finding and estimating the normalized eigenfunction pairs (Φ_k, Ψ_k) and associated eigenvalues, Λ_k, for $\mathbb{A}_0$ is a complicated, but entirely standard [7], eigenvalue problem for a system of ordinary differential equations with boundary conditions of standard type. It can be seen, and the details will appear elsewhere, that the Λ_k asymptotically lie along the rays

$$\arg(z) = \pm(\pi/2 + \epsilon)$$

in the left half complex plane, where ϵ, satisfying $0 < \epsilon \leq \pi/2$, depends upon $\rho(x)$, $H(x)$, and $EI(x)$ only, not upon the boundary conditions specified via B_0. The eigenfunction pairs (Φ_k, Ψ_k) may be shown to form a Riesz basis for $\mathcal{H}$. It is possible that some finite number of the Λ_k have positive real part because $\mathbb{A}_o$ is not dissipative in general. The Riesz basis property of the (Φ_k, Ψ_k) together with the asymptotic location of the Λ_k are all that are required to show that $\mathbb{A}_0$ generates a holomorphic semigroup on $\mathcal{H}$ (cf. [14]).

Now let (φ_k, ψ_k) and λ_k denote the normalized eigenfunction pairs and associated eigenvalues for $\mathbb{A}$. It turns out that one can prove that

$$\lim_{|\lambda_k| \to \infty} (|\Lambda_k - \lambda_k|) = 0,$$

$$\sum_k \|(\Phi_k, \Psi_k) - (\varphi_k, \psi_k)\|^2 < \infty.$$

From the last relation, using a standard Hilbert space result (see [19, p. 208ff.]) it can be seen that the (φ_k, ψ_k) also form a Riesz basis for $\mathcal{H}$ so that $\mathbb{A}$ is seen also to generate a holomorphic semigroup on $\mathcal{H}$.

4.6. Spline/Finite Element Approximation.

According to the principle of virtual work [10], given $w(t)$, the trajectory of a mechanical system subject to holonomic constraints $H(w(t)) = 0$, with forces $f_i(t)$, $i = 1, 2, \cdots, r$, acting on the system, the work done by these forces in a small virtual displacement from $w(t)$, at any instant t, vanishes. More precisely, if v is a vector such that

$$\frac{\partial H}{\partial w}(w(t))\, v = 0$$

then

$$\sum_{i=1}^{r} f_i(t)^* v = 0. \tag{4.46}$$

It should be noted that the sum (4.46) must include *all* forces acting on the system: inertial, restoring, damping, exogenous, etc. If a damping force, linearly dependent on velocity,

$$g(t) = -G\dot{w}(t),$$

where G is a nonnegative selfadjoint operator (matrix in the finite-dimensional case) is introduced into an otherwise conservative system whose kinetic energy is $\frac{1}{2}\dot{w}(t)^* M\dot{w}(t)$ and potential energy is $\frac{1}{2}w(t)^* Vw(t)$, M and V also nonnegative selfadjoint, M positive definite, (4.46) may be expressed as

$$(-M\ddot{w}(t) - G\dot{w}(t) - Vw(t))^* v = 0 \tag{4.47}$$

where $-M\ddot{w}$ is the inertial force, $-G\dot{w}$ is the damping force, and $-Vw$ is the restoring force for the system. Replacing v by $\dot{w}$ (which must be an admissible value for v), we obtain

$$\frac{d\mathcal{E}}{dt}(w(t), \dot{w}(t)) = -\dot{w}(t)^* G\dot{w}(t) \tag{4.48}$$

where $\mathcal{E}$ is the energy

$$\mathcal{E}(w(t)), \dot{w}(t)) = \tfrac{1}{2}\left(\dot{w}(t)^* M\dot{w}(t) + w(t)^* Vw(t)\right). \tag{4.49}$$

Formally, the derivation of the equation of motion,

$$M\ddot{w}(t) + G\dot{w}(t) + Vw(t) = 0, \tag{4.50}$$

is carried out via (4.47); in practice it is usually inferred from (4.48) after a quadratic form $-\dot{w}(t)^* G\dot{w}(t)$ expressing the energy loss through damping is hypothesized. In spatially distributed systems, the forms in which G, V appear in (4.47) may be different from the forms in which they appear in (4.48), (4.50) due to transformation via integration by parts (use of the divergence theorem), appearance of boundary terms, etc. For example, (4.48) might be expressed as a quadratic form in $\partial^2 w/\partial x\, \partial t$, but application of (4.47) requires that v should be an admissible velocity—not a partial derivative of such a velocity, necessitating a transformation by integration of parts.

To implement a spline, or general finite element, approximation procedure for the damped beam equation we begin with a finite-dimensional vector Φ of coordinate functions

$$\Phi = \begin{pmatrix} \varphi_1 \\ \vdots \\ \varphi_n \end{pmatrix}. \tag{4.51}$$

We may, optionally, assume that

$$\langle \beta, \varphi_i \rangle = 0, \qquad i = 1, 2, \cdots, n, \quad (\beta, 0) \in C,$$

so that the φ_i satisfy the kinematic constraints from the outset, or these constraints may be imposed in a later step to be described. For beam applications the basis functions φ_i, $i = 1, 2, \cdots, n$, must be such that, with

$$w(x,t) = \sum_{i=1}^{n} w_i(t)\varphi_i(x) = \Phi(\xi)^* w(t), \tag{4.52}$$

$$\frac{\partial w}{\partial t}(\xi,t) = \sum_{i=1}^{n} \dot{w}_i(t)\varphi_i(\xi) = \Phi(\xi)^* \dot{w}(t), \tag{4.53}$$

$$\frac{\partial^2 w}{\partial t\, \partial x}(\xi,t) = \sum_{i=1}^{n} \dot{w}_i(t)\varphi_i'(\xi) = \Phi'(\xi)^* \dot{w}(t), \tag{4.54}$$

are well-defined quantities at points ξ where external lateral forces and torques are to be applied.

Use of the representation (4.52), with concomitant expression of $\partial w/\partial t$, etc., as in (4.53), (4.54) implies movement of the beam is subject to certain holonomic constraints, infinitely many of them, in principle; the net effect of which is to allow the system to move in the finite-dimensional subspace spanned by the coordinate functions φ_i, $i = 1, 2, \cdots, n$. With such representation of $w(x,t)$, the work-energy relation with external lateral force σ and torque τ applied at the point $x = \xi$ may be seen to be

$$\begin{aligned} &\frac{d}{dt}\left(\dot{w}(t)^* M w(t) + w(t)^* V w(t)\right) - \dot{w}(t)^* G \dot{w}(t) \\ &\quad = \sigma \Phi(\xi)^* \dot{w}(t) + \tau \Phi'(\xi)^* \dot{w}(t). \end{aligned} \tag{4.55}$$

The functions φ_i are often chosen to be independent basis functions but this is not always convenient at the outset. If the φ_i are not independent, there will be a set of r independent linear equations satisfied by the accompanying coefficients, which we may represent in the form

$$P w(t) = 0, \tag{4.56}$$

where P is an $r \times n$ matrix, $r \leq n - m < n$. Let the $m \times n$ matrix Q be such that $\binom{P}{Q}$ is nonsingular and define $Qw = z$. We then have

$$\begin{pmatrix} P \\ Q \end{pmatrix} w(t) = \begin{pmatrix} 0 \\ z(t) \end{pmatrix}.$$

Writing $\binom{P}{Q}^{-1} = (T, U)$, we find that $w(t) = U z(t)$ and, with

$$\tilde{M} \equiv U^* M U, \quad \tilde{V} \equiv U^* V U, \quad \tilde{G} \equiv U^* G U,$$

we have, in place of (4.55),

$$\frac{d}{dt}\left(\dot{z}(t)^*\tilde{M}\dot{z}(t) + z(t)^*\tilde{V}z(t)\right) - \dot{z}(t)^*\tilde{G}\dot{z}(t) \\ = \sigma(t)\Phi(\xi)^*U\dot{z}(t) + \tau(t)\Phi'(\xi)^*U\dot{z}(t), \tag{4.57}$$

leading to the equations

$$\tilde{M}\ddot{z}(t) + \tilde{G}\dot{z}(t) + \tilde{V}z(t) = \sigma(t)U^*\Phi(\xi) + \tau(t)U^*\Phi'(\xi) \tag{4.58}$$

as the equations of motion for the system.

The kinematic boundary conditions, corresponding to C as described above, may be imposed via (4.56) if desired. The other boundary conditions, corresponding to $B - C$, are never imposed explicitly; they come about as particular cases of the dynamical equations in (4.58).

If the coordinates to be retained are a subset of the components of w, we can write $w^* = (\tilde{w}^*, \hat{w}^*)$, where $\tilde{w}$ comprises the components to be discarded and $\hat{w}$ the components to be retained. Then P is partitioned accordingly: $P = (\tilde{P}, \hat{P})$, and we can assume that the $r \times r$ matrix $\tilde{P}$ is nonsingular. Then, with $m = n - r$,

$$\tilde{w} = -(\tilde{P})^{-1}\hat{P}\hat{w}, \qquad w = \begin{pmatrix} -(\tilde{P})^{-1}\hat{P} \\ I_m \end{pmatrix}\hat{w} \equiv U\hat{w}$$

and we obtain (4.58) again, replacing $z(t)$ by $\hat{w}(t)$. Computationally this is easier because $\tilde{P}$ is $r \times r$ while $\binom{P}{Q}$ is $n \times n$.

In actual computational work, it is preferable to define the energy and dissipation forms in the continuous context and then approximate them by quadratic forms on finite-dimensional spaces. Application of (4.56), (4.57) to the approximate energy and dissipation forms then leads to the finite-dimensional approximating equations of motion in the form (4.58) with symmetric matrices $\tilde{M}$, $\tilde{G}$, and $\tilde{V}$. This procedure is much easier and more reliable then first deriving the equations of motion as partial differential equations in the continuous case and then proceeding to solve those equations numerically, taking into account all of the boundary conditions. In fact, by the recommended procedure, only the kinematic boundary conditions are actually enforced explicitly. The example of the next section should be helpful in clarifying these matters.

4.7. Approximation and Simulation for the Cantilever Beam System.

Let the interval $\mathcal{I} = [0, L]$ be divided into m subintervals

$$I_k = \left[(k-1)\frac{L}{m}, k\frac{L}{m}\right] \equiv [x_{k-1}, x_k] = [x_{k-1}, x_{k-1} + h], \qquad h = \frac{L}{m}, \\ k = 1, 2, \cdots$$

and consider general quadratic functions on the intervals I_k:

$$q_k(x) = a_k + b_k(x - x_{k-1}) + c_k(x - x_{k-1})^2.$$

The functions defined on I_k by 1, $x - x_{k-1}$, $(x - x_{k-1})^2$, and extended by 0 to the rest of $\mathcal{I}$ correspond to the functions φ_i, $i = 1, 2, \cdots, n$, $n = 3m$, discussed in the previous section. With

$$a(t)^* = (a_1(t), \cdots, a_m(t)), \qquad b(t)^* = (b_1, \cdots, b_m(t)),$$
$$c(t)^* = (c_1(t), \cdots, c_m(t)),$$
$$\Phi^*(x) = \left(\chi_{I_1}, \cdots, \chi_{I_m}, x'\chi_{I_1}, \cdots, (x - x_{m-1})'\chi_{I_m}, x^{2'}\chi_{I_1}, \cdots, (x - x_{m-1})^{2'}\chi_{I_m}\right)$$

and $w(t)^* = (a(t)^*, b(t)^*, c(t)^*)$, we can define a function $w(x,t)$, piecewise quadratic in x, with possible discontinuities at the points x_k, $k = 1, 2, \cdots, m-1$, by

$$w(x,t) = \Phi(x)^* w(t).$$

We proceed to define the energy and dissipation forms in terms of these functions, ignoring the discontinuities for the present.

Assuming the beam to have mass density ρ, the kinetic energy is

$$\begin{aligned}
\mathcal{T} &= \frac{\rho}{2}\int_0^L \left(\frac{\partial w}{\partial t}(x,t)\right)^2 dx \\
&= \frac{\rho}{2}\sum_{k=1}^{m}\int_{x_{k-1}}^{x_k} \left(\dot a_k(t) + \dot b_k(t)(x - x_{k-1}) + \dot c_k(t)(x - x_{k-1})^2\right) dx \\
&= \frac{\rho}{2}\left(h\dot a_k(t)^2 + \frac{h^3}{3}b_k(t)^2 + \frac{h^5}{5}\dot c_k(t) + h^2\dot a_k(t)\dot b_k(t)\right. \\
&\qquad \left. + \frac{2}{3}h^3\dot a_k(t)\dot c_k(t) + \frac{1}{2}h^4\dot b_k(t)\dot c_k(t)\right) = \frac{\rho}{2}\dot w(t)^* M \dot w(t),
\end{aligned}$$

from which we see that

$$M = \begin{pmatrix} hI_m & \frac{1}{2}h^2 I_m & \frac{1}{3}h^3 I_m \\ \frac{1}{2}h^2 I_m & \frac{1}{3}h^3 I_m & \frac{1}{4}h^4 I_m \\ \frac{1}{3}h^3 I_m & \frac{1}{4}h^4 I_m & \frac{1}{5}h^5 I_m \end{pmatrix}.$$

Here I_m denotes the $m \times m$ identity matrix.

Next we consider the potential energy, which for constant bending modulus EI, is

$$\frac{1}{2}EI\sum_{k=1}^{m}\int_{x_{k-1}}^{x_k}\left(\frac{\partial^2 w}{\partial x^2}(x,t)\right)^2 dx = \frac{1}{2}EI\sum_{k=1}^{m}\int_{x_{k-1}}^{x_k} 4\dot{c}_k(t)^2\,dx$$
$$= 2EIh\dot{c}_k(t)^2$$

so that

$$V = \begin{pmatrix} 0 & 0 & 0 \\ 0 & 0 & 0 \\ 0 & 0 & 4EIhI_m \end{pmatrix}.$$

Passing now to the dissipation form, we first of all take the interaction kernel $h(x,\xi)$ to have the form $\gamma h(x-\xi)$ where, assuming that m will always be a multiple of a fixed positive integer μ,

$$h(\eta) = 1, \quad |\eta| \le L/\mu, \quad h(\eta) = 0 \text{ otherwise.}$$

We define $\nu = \nu(m) = m/\mu$ for all values of m used. For the case $m = 9$, $\nu = \mu = 3$, the support of $h(x-\xi)$ is the shaded region shown in Fig. 4.7.

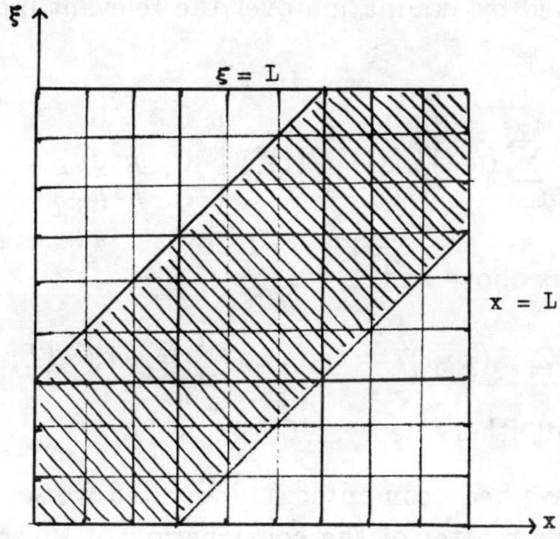

FIG. 4.7

Let

$$S_k^\ell = \{(x,\xi)|x_{k-1} \le x \le x_k, \quad x_{\ell-1} \le \xi \le x_\ell\}.$$

Then in S_k^ℓ we have

$$\frac{\partial^2 w}{\partial t\,\partial x}(x,t) - \frac{\partial^2 w}{\partial t\,\partial x}(\xi,t)$$
$$= \dot{b}_k(t) + 2\dot{c}_k(t)(x - x_{k-1}) - \dot{b}_\ell(t) - 2\dot{c}_\ell(t)(\xi - \xi_{\ell-1}).$$

For $|k-\ell| < \nu$ we square this quantity and integrate over S_k^ℓ; then sum over the indicated values of k,ℓ to obtain the quadratic form

$$\dot{w}(t)^* G_0 w(t) = 2\gamma \sum_{k,\ell} (\dot{a}(t)^*, \dot{b}(t)^*, \dot{c}(t)^*) \begin{pmatrix} 0 & 0 & 0 \\ 0 & h^2 J_{k,\ell} & h^3 J_{k,\ell} \\ 0 & h^3 J_{k,\ell} & h^4 K_{k,\ell} \end{pmatrix} \begin{pmatrix} \dot{a}(t) \\ \dot{b}(t) \\ \dot{c}(t) \end{pmatrix}$$

where

$$(J_{k,\ell})_k^k = 1, \quad (J_{k,\ell})_\ell^\ell = 1, \quad (J_{k,\ell})_\ell^k = -1, \quad (J_{k,\ell})_k^\ell = -1,$$

$$(J_{k,\ell})_j^i = 0 \quad \text{otherwise;}$$

$$(K_{k,\ell})_k^k = \frac{4}{3}, \quad (K_{k,\ell})_\ell^\ell = \frac{4}{3}, \quad (K_{k,\ell})_\ell^k = -1, \quad (K_{k,\ell})_k^\ell = -1,$$

$$(K_{k,\ell})_j^i = 0 \quad \text{otherwise.}$$

Additionally, we must treat the upper and lower borders of the region of support of $h(x - \xi)$, consisting of triangles $\tilde{T}_k$, $\hat{T}_k$, $k = 1, 2, \cdots, M - \nu$ as shown in Fig. 4.7. After integration of the same function as shown earlier over these regions, followed by summation over the relevant indices, we obtain

$$\dot{w}(t)^* G_1 \dot{w}(t) = 2\gamma \sum_{k=1}^{m-\nu} (\dot{a}(t)^*, \dot{b}(t)^*, \dot{c}(t)^*) \begin{pmatrix} 0 & 0 & 0 \\ 0 & h^2 J_{k,\ell} & h^3 K_{k,\ell} \\ 0 & h^3 K_{k,\ell} & h^4 L_{k,\ell} \end{pmatrix} \begin{pmatrix} \dot{a}(t) \\ \dot{b}(t) \\ \dot{c}(t) \end{pmatrix}$$

with $J_{k,\ell}$ and $K_{k,\ell}$ as above and

$$(L_{k,\ell})_k^k = 2, \quad (L_{k,\ell})_\ell^\ell = 2, \quad (L_{k,\ell})_\ell^k = -1, \quad (L_{k,\ell})_k^\ell = -1,$$

$$(L_{k,\ell})_j^i = 0 \text{ otherwise;}$$

After G_0 and G_1 have been computed as indicated we set $G = G_0 + G_1$.

Next there is the matter of the construction of the reduction matrix U. From the representation on I_k

$$w(x,t) = a_k(t) + b_k(t)(x - x_{k-1}) + c_k(t)(x - x_{k-1})^2,$$

$$\frac{\partial w}{\partial x}(x,t) = b_k(t) + 2c_k(t)(x - x_{k-1}),$$

we see that $w(x,t)$ is piecewise C^2 on $\mathcal{I}$ just in case, for $k = 1, 2, \cdots, m - 1$ we have

$$a_{k+1}(t) = a_k(t) + b_k(t)h + c_k(t)h^2,$$
$$b_{k+1}(t) = b_k(t) + 2c_k(t)h.$$

For the cantilever case of our example we have, in addition,

$$a_1(t) = 0, \qquad b_1(t) = 0.$$

Accordingly, we have the following linear relationships holding between the coefficients

$$\left(\underbrace{\begin{matrix} 1 & 0 & 0\cdots 0 & 0 & 0 & 0 & 0\cdots 0 & 0 \\ -1 & 1 & 0\cdots 0 & 0 & 0 & -h & 0\cdots 0 & 0 \\ 0 & -1 & 1\cdots 0 & 0 & 0 & -h & 0\cdots 0 & 0 \\ \vdots & \vdots & \vdots & \vdots & \vdots & \vdots & \vdots & \vdots \\ 0 & 0 & 0\quad -1 & 1 & 0 & 0 & 0\cdots -h & 0 \\ 0 & 0 & 0\cdots 0 & 0 & 1 & 0 & 0\cdots 0 & 0 \\ 0 & 0 & 0\cdots 0 & 0 & -1 & 1 & 0\cdots 0 & 0 \\ 0 & 0 & 0\cdots 0 & 0 & 0 & -1 & 1\cdots 0 & 0 \\ \vdots & \vdots & \vdots & \vdots & \vdots & \vdots & \vdots & \vdots \\ 0 & 0 & 0\quad 0 & 0 & 0 & 0 & 0\cdots -1 & 1 \end{matrix}}_{\tilde{P}} \quad \underbrace{\begin{matrix} 0 & 0 & 0\cdots 0 & 0 \\ -h^2 & 0 & 0\cdots 0 & 0 \\ 0 & -h^2 & 0\cdots 0 & 0 \\ \vdots & \vdots & \vdots & \vdots \\ 0 & 0 & 0\cdots -h^2 & 0 \\ 0 & 0 & 0\cdots 0 & 0 \\ -2h & 0 & 0\cdots 0 & 0 \\ 0 & -2h & 0\cdots 0 & 0 \\ \vdots & \vdots & \vdots & \vdots \\ 0 & 0 & 0\cdots -2h & 0 \end{matrix}}_{\hat{P}}\right) \begin{pmatrix} a_1 \\ \vdots \\ a_m \\ b_1 \\ \vdots \\ b_m \\ c_1 \\ \vdots \\ c_m \end{pmatrix}$$

$$= 0.$$

Let R be the matrix with

$$R^i_j = -1, \quad j = i + 1, \quad r^i_j = 0 \text{ otherwise}.$$

Then, with $Q = I + R$ we have

$$\tilde{P} = \begin{pmatrix} I + R & hR \\ 0 & I + R \end{pmatrix} = \begin{pmatrix} Q & hR \\ 0 & Q \end{pmatrix}$$

is readily inverted, giving

$$\left(\tilde{P}\right)^{-1} = \begin{pmatrix} Q^{-1} & -h(Q^{-1} - Q^{-2}) \\ 0 & Q^{-1} \end{pmatrix}.$$

Since

$$\hat{P} = \begin{pmatrix} h^2 R \\ 2hR \end{pmatrix} = \begin{pmatrix} h^2(Q - I_m) \\ 2h(Q - I_m) \end{pmatrix},$$

we obtain, as the reduction matrix

$$U = \begin{pmatrix} -\tilde{P}^{-1}\hat{P} \\ I_m \end{pmatrix} = \begin{pmatrix} h^2 I_m - 3h^2 Q^{-1} + 2h^2 Q^{-2} \\ 2h(Q^{-1} - I_m) \\ I_m \end{pmatrix}.$$

Then a further easy computation shows that

$$Q^{-1} = \begin{pmatrix} 1 & 0 & 0 & \cdots & 0 \\ 1 & 1 & 0 & \cdots & 0 \\ 1 & 1 & 1 & \cdots & 0 \\ \vdots & \vdots & \vdots & & \vdots \\ 1 & 1 & 1 & \cdots & 1 \end{pmatrix}, \qquad Q^{-2} = \begin{pmatrix} 1 & 0 & 0 & \cdots & 0 \\ 2 & 1 & 0 & \cdots & 0 \\ 3 & 2 & 1 & \cdots & 0 \\ \vdots & \vdots & \vdots & & \vdots \\ m & m-1 & m-2 & \cdots & 1 \end{pmatrix}.$$

We see therefore that the reduction matrix U can be computed explicitly without having to resort to numerical matrix inversion in this example. Once this has been done the matrices $\tilde{M}$, $\tilde{G}$, and $\tilde{V}$ of (4.57) can be calculated readily and, using a Cholesky decomposition to write $\tilde{M}$ in the form $\mathcal{L}\mathcal{L}^*$, with $\mathcal{L}$ lower triangular, the equation (4.58) can be pre- and post-multiplied by $\mathcal{L}^{-1}$ and $(\mathcal{L}^*)^{-1}$, respectively, to yield an equation for which the corresponding first-order system matrix is

$$\mathcal{A} = \begin{pmatrix} 0 & I_m \\ -\mathcal{L}^{-1}\tilde{V}(\mathcal{L}^*)^{-1} & -\mathcal{L}^{-1}\tilde{G}(\mathcal{L}^*)^{-1} \end{pmatrix}.$$

We have calculated this matrix, and subsequently its eigenvalues, for a cantilever beam with $\rho = 1$, $\gamma = .01$, $EI = 100$, and length $L = 10$. The interaction kernel $h(x - \xi)$ was taken to be constant with support restricted to $|x - \xi| \le L/2$. In Figs. 4.8(a) and (b) we show the eigenvalue patterns in the upper left half plane for $m = 14, 16$, and 24 (not all of the eigenvalues are shown in the last case). Figure 4.8(b) magnifies the lower right-hand corner of Fig. 4.8(a) in order to show the pattern of the eigenvalues corresponding to low frequencies more clearly. Variational considerations account for the decrease in the natural frequencies as M gets larger but the increasing negative real parts of eigenvalues, or exponential damping rates, came as somewhat of a surprise. This property, if it can be substantiated in general, would be quite desirable for applications because it would mean that damping rates are systematically underestimated in approximations of this sort, providing a safety factor in control applications. From Fig. 4.8 it is seen that the quadratic dependence of the damping rate on frequency at the low end of the spectrum has only the most limited range.

4.8. Some Theory Related to Laboratory Experimentation.

The foregoing results from mathematical simulation can be, and have been, compared with related data taken from elastic beam experiments in a laboratory setting. This, of course, is nothing new by itself (see, e.g., the articles [3], [4] and accompanying references) but perhaps we can claim that we are in a position to look at such data in a different light than most of our predecessors as the author is fortunate to have access to extensive electronic equipment for collecting and analyzing data. Before describing the results of this experimentation program and the relationship of these results to the mathematical simulation discussed above, we digress briefly to describe the experimental context and some of the inherent problems attendant upon the observation of vibrating beams in the laboratory.

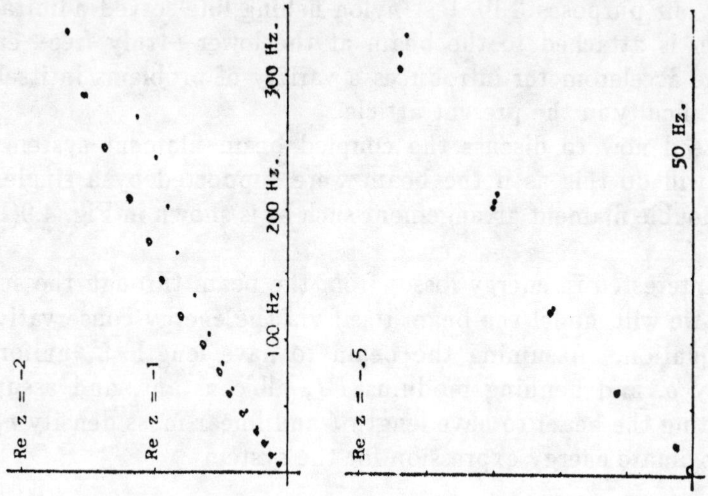

FIG. 4.8. *Location of eigenvalues for* $M = 14, 16, 24$. *Eigenvalues move to the left and down as* M *increases.*

The laboratory beam configurations easiest to realize, at least approximately, are those involving either clamped ends or free ends. The results which we cite in §4.9 were obtained in the clamped/free (cantilever) setting and in a "pseudo" free/free configuration. That the task of realizing a good approximation to a clamped beam endpoint is not at all trivial will be commented on at some length later in this section. The free/free case is qualified by the word "pseudo" because, in fact, the beam is actually suspended vertically by a very fine nylon thread. The two experimental configurations are illustrated in Fig. 4.9(a), (b).

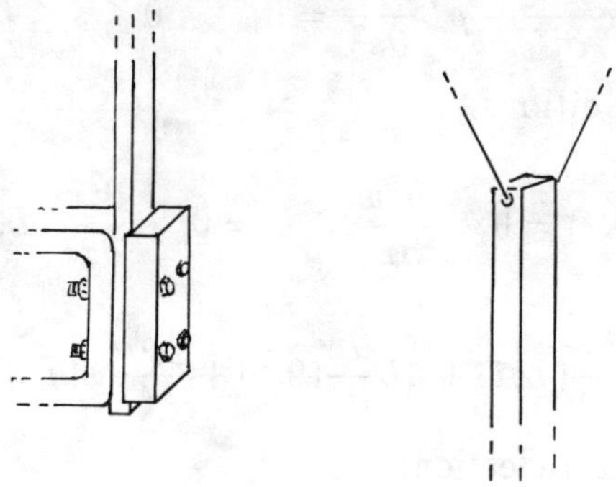

FIG. 4.9

Let us begin by discussing this pseudo free/free case; from now on we simply refer to it as the pseudo free configuration. To obtain a reasonable facsimile of the free beam, whose actual realization requires a gravity-free environment achievable only in space or free fall conditions with accompanying difficulties as regards excitation and measurement, we suspend an elastic beam from a suitable support, preferably a high one, by a long, very fine, nylon filament; for our purposes 3 lb. test nylon fishing line served admirably. The accelerometer is attached to the beam at the lower—truly free—end. Just attaching the accelerometer introduces a variety of problems in itself but we avoid that difficulty in the present article.

We proceed now to discuss the coupled beam–filament system in some detail. We will do this as if the beam were supported by a single filament although a double filament arrangement such as is shown in Fig. 4.9(b) is used in practice.

We are interested in energy losses from the beam through the supporting filament, so we will model the beam itself via the energy conservative Euler–Bernoulli equation. Assuming the beam to have length L, uniform linear mass density ρ, and bending modulus EI, all constant, and assuming the cord supporting the beam to have length ℓ and linear mass density r, we have as the approximate energy expression for the system

$$\frac{1}{2}\int_0^L \left[\rho\left(\frac{\partial w}{\partial t}\right)^2 + EI\left(\frac{\partial^2 w}{\partial x^2}\right)^2\right] dx + \frac{1}{2} g\rho L\left(\frac{v(0,t)}{\ell}\right)^2$$
$$+\frac{1}{2}\int_0^\ell \left[r\left(\frac{\partial v}{\partial t}\right)^2 + \rho L\left(\frac{\partial v}{\partial s}\right)^2\right] ds,$$

with the constraint $w(L,t) = v(0,t)$. Here w is the beam deflection as a function of x, and v is the string deflection as a function of s. From this energy form one readily derives the equations

$$\rho\frac{\partial^2 w}{\partial t^2} + EI\frac{\partial^4 w}{\partial x^4} = 0, \qquad 0 \le x \le L, \tag{4.59}$$

$$r\frac{\partial^2 w}{\partial t^2} - \rho L\frac{\partial^2 w}{\partial s^2} = 0, \qquad 0 \le s \le \ell, \tag{4.60}$$

and the boundary conditions

$$\frac{\partial^2 w}{\partial x^2}(0,t) = 0, \quad \frac{\partial^3 w}{\partial x^3}(0,t) = 0, \quad \frac{\partial^2 w}{\partial x^2}(L,t) = 0, \tag{4.61}$$

$$EI\frac{\partial^3 w}{\partial x^3}(L,t) + \rho L\frac{\partial v}{\partial s}(0,t) + \frac{g\rho L}{\ell}w(L,t) = 0. \tag{4.62}$$

At $s = \ell$ we assume no reflection:

$$\frac{\partial v}{\partial t}(\ell,t) + \left(\frac{\rho L}{r}\right)^{1/2}\frac{\partial v}{\partial s}(\ell,t) = 0. \tag{4.63}$$

It is not supposed that the last condition is completely realistic. It simply replaces more complicated assumptions ensuring that very little, if any, energy entering the supporting filament, or string, is reflected back to the beam. We are assuming that waves in the string move only in the direction from $s = 0$ to $s = \ell$. Then, in fact, (4.63) is valid with ℓ replaced by s, $0 \le s \le \ell$; in particular, it is true at $s = 0$. Using this in (4.62) along with the constraint $w(L,t) = v(0,t)$ we arrive at

$$EI\frac{\partial^3 w}{\partial x^3}(L,t) + \left[\frac{g\rho L}{\ell} - (r\rho L)^{1/2}\right]\frac{\partial w}{\partial t}(L,t) = 0. \tag{4.64}$$

Since we assume ℓ large and r small,

$$\tau \equiv (r\rho L)^{1/2} - \frac{g\rho L}{\ell}$$

is a small positive number and the boundary condition (4.64) becomes

$$EI\frac{\partial^3 w}{\partial x^3}(L,t) - \tau\frac{\partial w}{\partial t}(L,t) = 0. \tag{4.65}$$

If we define

$$a = (\rho/EI)^{1/4}$$

and assume solutions of the form

$$w(x,t) = e^{i\omega^2 t}w(x)$$

with ω complex, the resulting eigenfunction equation is

$$w^{(iv)} - (a\omega)^4 w = 0. \tag{4.66}$$

Applying the boundary conditions (4.63) and (4.65) to the general solution of (4.66), and setting

$$\lambda = a\omega, \qquad \sigma = \tau/a^2,$$

we obtain the determinantal condition

$$\det\begin{pmatrix} -\cos\lambda L + \cosh\lambda L & \sin\lambda L + \sinh\lambda L \\ \lambda\sin\lambda L + \lambda\sinh\lambda L & -\lambda\cos\lambda L + \lambda\cosh\lambda L \\ -\sigma i\cos\lambda L - \sigma i\cosh\lambda L & -\sigma i\sin\lambda L - \sigma i\sinh\lambda L \end{pmatrix} = 0,$$

which simplifies to

$$2\lambda - 2\lambda\cos\lambda L\cosh\lambda L + 2\sigma i\cos\lambda L\sinh\lambda L - 2\sigma i\sin\lambda L\cosh\lambda L = 0.$$

If we let

$$\lambda = \left(\frac{2k-1}{2}\right)\frac{\pi}{L} + \delta\lambda$$

and substitute into this equation, we find that, to first order in $\delta\lambda$ we have

$$\delta\lambda\left[\cosh\left(\frac{2k-1}{2}\right)\pi - \frac{2\sigma iL}{2(k-1)\pi}\sinh\left(\frac{(2k-1)\pi}{2}\right)\right]$$
$$\simeq (-1)^k + \frac{2\sigma i}{(2k-1)\pi}\cosh\left(\frac{(2k-1)\pi}{2}\right),$$

and from this it is easy to see that, to first order in $1/k$,

$$\delta\lambda \simeq \frac{2\sigma i}{(2k-1)\pi}.$$

Then, to zero order in $1/k$,

$$i\omega^2 = i\frac{\lambda^2}{a^2} = 1\frac{(2k-1)^2\pi^2}{4a^2L^2} - \frac{2\sigma}{a^2L}.$$

Asymptotically, then, the damping of the kth mode of the beam due to the presence of the supporting filament is uniform and the damping exponent

$$\frac{2\sigma}{a^2L} = \frac{2EI\tau}{\rho L} = \frac{2EI}{\rho L}\left((r\rho L)^{1/2} - \frac{g\rho L}{\ell}\right)$$

tends to zero as $r \to 0$ and $\ell \to \infty$, i.e., as the supporting cord becomes very light and very long. We judge, therefore, that this type of support mechanism will not lead to damping rates comparable at high frequencies to those arising from internal sources in our laboratory experiments.

A somewhat similar analysis shows that losses due to excitation of acoustic waves in the surrounding atmosphere by the high frequency vibrations of the beam will also be asymptotically uniform and thus not likely to be confused with internal damping effects.

It has been recognized by various experimenters for some time that clamped beam experiments are subject to substantial degrees of interference from the interaction between the beam and the nearly, but not quite, rigid clamping device. In order to have some means for evaluating the likely degree of error in damping experiments under these circumstances, we ask the reader to consider with us a clamping arrangement in which the beam to be tested is, as shown in Fig. 4.10, the relatively thin beam lying in the region $x > 0$, except for a portion embedded in the more massive beam which occupies the region $x < 0$ and which serves as the clamping device; stiff but not entirely rigid—the unintended result of most of our laboratory clamping attempts.

We denote lateral displacement of the thin beam by $w(x,t)$, $x \geq 0$, and lateral displacements of the thick beam by $W(x,t)$, $x \leq 0$. We will study "monochromatic," i.e., single frequency, solutions of the combined system under the assumption that "outgoing" waves, moving toward $-\infty$, in the clamping beam are eventually absorbed and not reflected back to $x = 0$.

Adopting complex notation, we represent the displacement of the thin beam by

$$w(x,t) = \beta e^{i(\omega^2 t - \alpha\omega x)} + \gamma e^{i(\omega^2 t + \alpha\omega x)}, \tag{4.67}$$

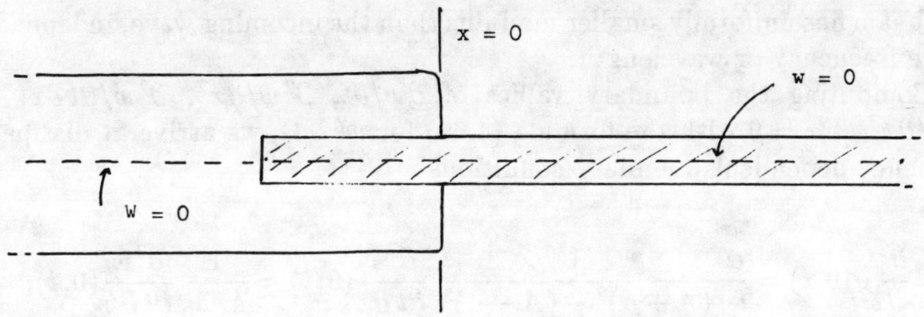

FIG. 4.10

where $\alpha^4 = (\rho/\epsilon\iota)$, ρ and $\epsilon\iota$ being the linear mass density and bending modulus, respectively, of the thin beam. The first term represents waves moving outward from $x = 0$ to infinity while the second term represents waves moving inward toward $x = 0$ from infinity. Because of our assumption that all waves in the thick beam originate at $x = 0$ and are not reflected back subsequently, we represent the thick beam displacement by

$$W(x,t) = \delta e^{i(\omega^2 t + A\omega x)}, \tag{4.68}$$

where $A^4 = (R/EI)$, and R and EI are the linear mass density and bending modulus, respectively, of the heavy beam. It is easy to see that the condition that no work should be done on the combined system at $x = 0$ is

$$EI\frac{\partial^2 W}{\partial x^2}(0,t) - \epsilon\iota\frac{\partial^2 w}{\partial x^2}(0,t) = 0,$$

$$EI\frac{\partial^3 W}{\partial x^3}(0,t) - \epsilon\iota\frac{\partial^3 w}{\partial x^3}(0,t) = 0,$$

where EI and $\epsilon\iota$ have been defined above. Applied to (4.67) and (4.68), these yield

$$\begin{pmatrix} \epsilon\iota\alpha^2\omega^2 & \epsilon\iota\alpha^2\omega^2 & -EIA^2\omega^2 \\ i\epsilon\iota\alpha^3\omega^3 & -i\epsilon\iota\alpha^3\omega^3 & EIA^3\omega^3 \end{pmatrix} \begin{pmatrix} \beta \\ \gamma \\ \delta \end{pmatrix} = \begin{pmatrix} 0 \\ 0 \end{pmatrix}.$$

Normalizing γ to 1 we find that

$$\delta = \frac{2\epsilon\iota\alpha^3}{EIA^2(A+\alpha)}, \qquad \beta = \frac{A-\alpha}{A+\alpha}.$$

It is convenient to renormalize β to $A - \alpha$, γ to $A + \alpha$ so that

$$w(x,t) = (A-\alpha)e^{i(\omega^2 t - \alpha\omega x)} + (A+\alpha)e^{i(\omega^2 t + \alpha\omega x)}. \tag{4.69}$$

If the two beams are of the same material, so that $E = \epsilon$, then, since ι becomes small more rapidly than ρ as the thickness of the thin beam decreases, we shall have $0 < A < \alpha$, $-1 < \beta < 0$ so that the outgoing, i.e., reflected, wave in the thin beam has uniformly smaller modulus than the incoming wave, independent of the frequency or wavelength.

Computing the boundary values of $\partial w/\partial t$, $\partial^2 w/\partial x^2$, $\partial^2 w/\partial t\,\partial x$, and $\partial^3 w/\partial x^3$ at $x = 0$ with the formula (4.69) for $w(x,t)$, we arrive at dissipative frequency dependent boundary conditions

$$\frac{\partial^2 w}{\partial x^2}(0,t) = \frac{\alpha}{\omega} \frac{1}{(A+\alpha)^2 - (A-\alpha)^2} \frac{\partial^2 w}{\partial t\,\partial x}(0,t) = \frac{1}{4A\omega} \frac{\partial^2 w}{\partial t\,\partial x}(0,t),$$

$$\frac{\partial^3 w}{\partial x^3}(0,t) = -\alpha^3\omega \frac{1}{(A+\alpha)^2 - (A-\alpha)^2} \frac{\partial w}{\partial t}(0,t) = \frac{-\alpha^2\omega}{4A} \frac{\partial w}{\partial t}(0,t).$$

It may be computed that the energy associated with the travelling wave of wavelength $2\pi/\alpha\omega$, call it $\epsilon(\omega,t)$, computed over a suitable finite, moving domain with zero as its left-hand boundary, in order to avoid ϵ being infinite, decays according to

$$\frac{d\mathcal{E}(\omega,t)}{dt} = -8A\rho\omega^5$$

while the energy itself, over a large domain as described, may be seen to be proportional to ω^4. We conclude, as a result, that the energy decay rate exponent must be directly proportional to the square root of the frequency, ω^2, under these circumstances. It is clear that such a damping relationship may seriously interfere with accurate measurement of the internal exponential damping rate, even though we expect the latter to be proportional to ω^2 rather than ω. While actual clamping mechanisms are not really thick beams as supposed here, this analysis is, we feel, sufficiently realistic to explain at least partially the very severe difficulties encountered in trying to measure the damping rates of natural modes of clamped beams in the laboratory.

4.9. Some Experimental Results.

In a laboratory experiment the beam is excited by striking it smartly in order to set it into vibration. A record such as the one shown in Fig. 4.11 is obtained; in this case the record exhibited is the graph of an input voltage to an HP5451C System Analyzer proportional to the free endpoint lateral velocity of a cantilever beam clamped at the other end.

The record initially shows a substantial high frequency component, as evidenced by the jagged character of the graph on the left-hand side. As the motion continues we observe that the curve becomes smoother, evidence that the high frequency component is becoming smaller in relation to the overall amplitude of the motion. In fact, the amplitude of the fundamental mode is not very different at the end of the time interval from what it was at the beginning; very little damping is observed at this low frequency.

To obtain a better appreciation of the quantitative factors involved in the evolution of the beam's motion, the record is divided into a number of successive segments of equal length. The (Fast) Fourier transform of each segment is computed and multiplied by its conjugate to yield the power

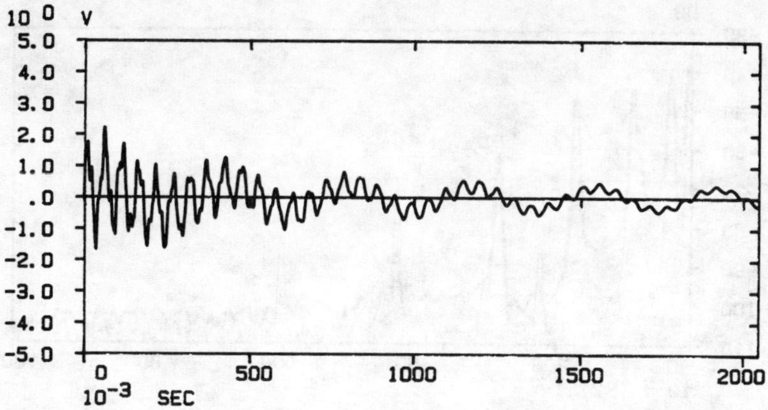

FIG. 4.11

spectrum for that interval. For most purposes it is desirable to form the logarithm of the power spectrum, which we henceforth call the logarithmic power spectrum. This is what is displayed in Fig. 4.12 (though for a different input record than the one shown in Fig. 4.13). Here two successive logarithmic power spectra are superimposed to yield the complete diagram. The peaks evident in each graph correspond to the frequencies of the natural modes of vibration. Since it is the logarithm which is displayed, the vertical gaps between peaks of successive spectra are proportional to the (negative) real part of the complex system eigenvalue associated with that mode of vibration while the horizontal position of the peak corresponds to the imaginary part of the eigenvalue, i.e., to the frequency of that mode.

Figures 4.12–4.14 display logarithmic power spectra from a "pseudo-free" boron-epoxy composite beam, a clamped wooden beam, and a "pseudo-free" metallic beam, respectively. In these figures we have added short horizontal bars to more clearly indicate the successive peak positions. The data is averaged over many experiments, using the same beam, recording data over an interval of the same length, but exciting the beam in different ways in order to adequately sample all modes. The approximately linear relationship between damping ratio and frequency is strikingly evident for the wooden and composite beams, less so, but still fairly convincing, for the metallic beam.

It might be natural as a result of this evidence of frequency proportional damping in elastic beams to suppose that all comparable elastic structures exhibit the same sort of relationship between damping and frequency. Indeed, this was the author's initial predisposition. It is here that continued experimentation in the laboratory teaches us not to trust our instincts too implicitly. Figure 4.15 shows successive logarithmic power spectra obtained from longitudinal vibrations of a steel rod, approximately one meter long and 1.5 cm. in diameter in its circular cross section. The accelerometer is mounted on one end and the other end is struck sharply with a heavy steel hammer. Through the first three modes the damping rates grow quite rapidly but then settle down to nearly constant rates, exhibiting no demonstrable dependence on frequency

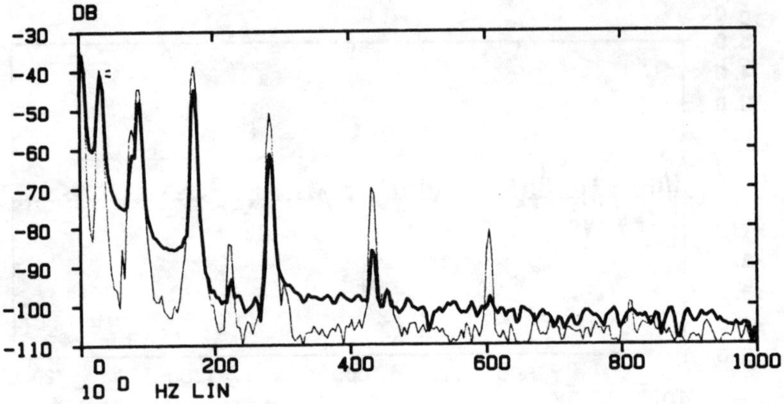

FIG. 4.12

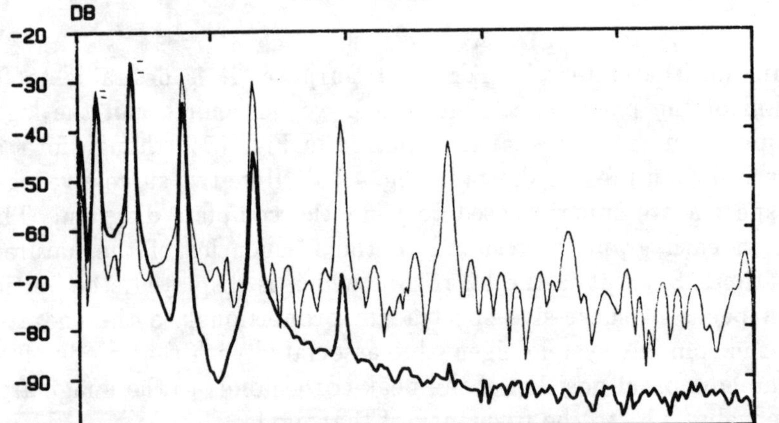

FIG. 4.13

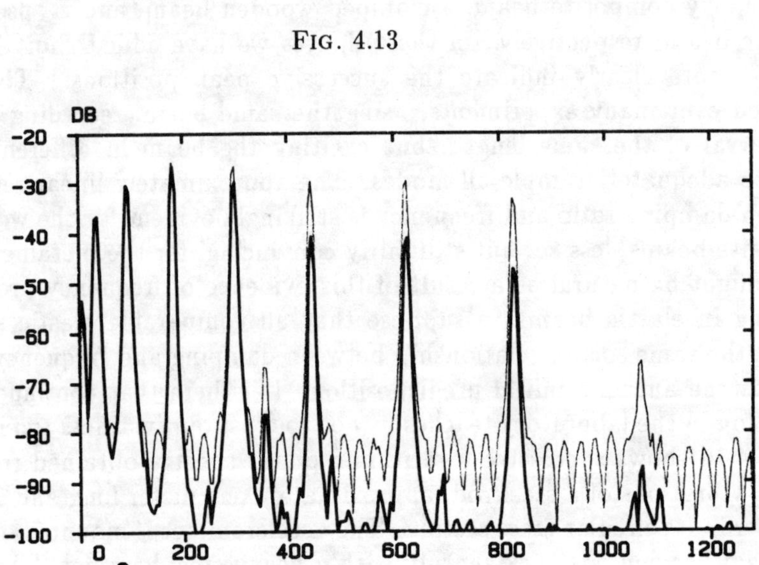

FIG. 4.14

from 10 khz. through 50 khz. This experiment has been repeated many times with much the same results in all cases.

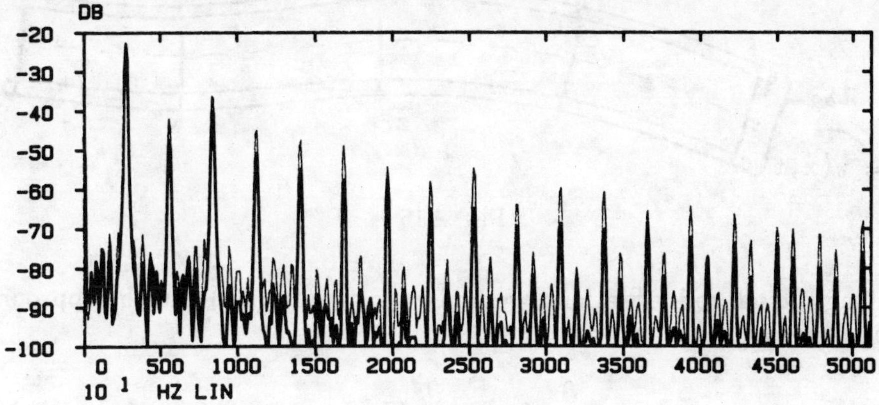

FIG. 4.15

In reality, this result is quite consistent with frequency proportional damping in the Euler–Bernoulli beam and with our earlier suppositions as to the source of the damping action. If the same assumptions are made in regard to the internal structure of the rod as were made for the beam, the resulting model for vibrations in the rod takes the form of the damped wave equation (constant coefficient case only shown here)

$$r\frac{\partial^2 z}{\partial t^2} + 2\gamma \int_0^L \ell(x-\xi)\left[\frac{\partial z}{\partial t}(x,t) - \frac{\partial z}{\partial t}(\xi,t)\right] d\xi - E\frac{\partial^2 z}{\partial x^2} = 0, \tag{4.70}$$

$z(x,t)$ denoting the longitudinal displacement of the rod element whose equilibrium center of mass is at x and $\ell(\eta)$ an interaction kernel of the same sort as $h(\eta)$ used earlier for the beam. Analysis similar to that used for the infinite beam in §4.4 shows that a damping spectrum of the type observed in Fig. 4.15, i.e., quadratic damping exponent versus frequency dependence for small ω and asymptotically constant as $\omega \to \infty$, results for the infinite rod and, correspondingly, similar discrete behavior is to be expected for a finite rod. So we may tentatively accept (4.70) as a viable model for longitudinal, internally damped vibrations of a rod.

With (4.70) so accepted, we now envision an idealized beam constructed, as shown in Fig. 4.16, from two such rods rigidly spaced a distance 2δ apart, with a mass-less, perfectly flexible, but inextensible central support structure forming the "backbone" of the system. Lateral displacements of the structure will be described in terms of the graph of the locus of the center line, $w = w(x,t)$, as shown in the figure.

Using the energy expression

$$\mathbb{E}\left(z, \frac{\partial z}{\partial t}\right) = \frac{1}{2}\int_0^L \left[r\left(\frac{\partial z}{\partial t}\right)^2 + E\left(\frac{\partial z}{\partial x}\right)^2\right] dx$$

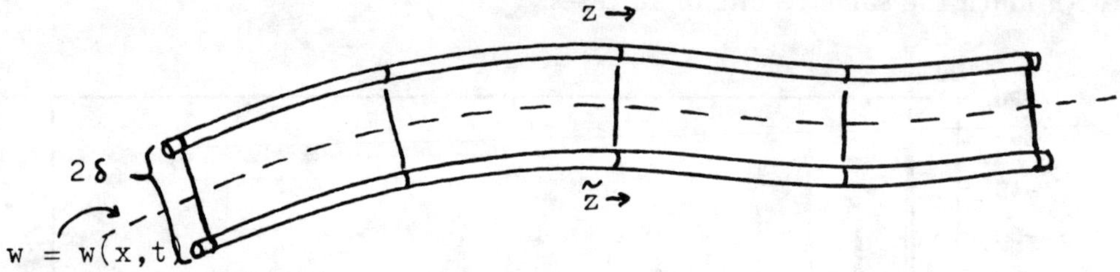

FIG. 4.16

for longitudinal rod vibrations, we find with use of (4.70) and the appropriate boundary conditions, e.g.,

$$\frac{\partial z}{\partial x}(0,t) = \frac{\partial z}{\partial x}(L,t) = 0 \tag{4.71}$$

in the case of free ends, that for smooth solutions we have

$$\frac{d\mathbb{E}}{dt}\left(z, \frac{\partial z}{\partial t}\right) = -\gamma \int_0^L \int_0^L \ell(x-\xi)\left[\frac{\partial z}{\partial t}(x,t) - \frac{\partial z}{\partial t}(\xi,t)\right]^2 d\xi\, dx. \tag{4.72}$$

Let $\hat{z}(x,t)$, $\tilde{z}(x,t)$ denote the longitudinal displacements of the upper and lower rods of the beam structure, respectively. Then we find, to first order in δ, that

$$\frac{\partial \hat{z}}{\partial x}(x,t) = -\delta \frac{\partial^2 w}{\partial x^2}(x,t), \qquad \frac{\partial \tilde{z}}{\partial x}(x,t) = \delta \frac{\partial^2 w}{\partial x^2}(x,t).$$

Letting

$$\mathcal{E}\left(w, \frac{\partial w}{\partial t}\right) = \mathbb{E}\left(\hat{z}, \frac{\partial \hat{z}}{\partial t}\right) + \mathbb{E}\left(\tilde{z}, \frac{\partial \tilde{z}}{\partial t}\right) + \int_0^L r\left(\frac{\partial w}{\partial t}\right)^2 dx \tag{4.73}$$

we have, with $\rho = 2r$, $I_\rho = \rho\delta$, $EI = 2\delta^2 E$,

$$\mathcal{E}\left(w, \frac{\partial w}{\partial t}\right) = \frac{1}{2}\int_0^L \left[\rho\left(\frac{\partial w}{\partial t}\right)^2 + I_\rho\left(\frac{\partial^2 w}{\partial t\,\partial x}\right)^2 + EI\left(\frac{\partial^2 w}{\partial x^2}\right)^2\right] dx \tag{4.74}$$

while dissipation form (4.72) becomes, with

$$h(x-\xi) = 2\delta^2 \ell(x-\xi),$$

$$\frac{d\mathcal{E}}{dt}\left(w, \frac{\partial w}{\partial t}\right) = -\gamma \int_0^L \int_0^L h(x-\xi)\left[\frac{\partial^2 w}{\partial t\,\partial x}(x,t) - \frac{\partial^2 w}{\partial t\,\partial x}(\xi,t)\right]^2 d\xi\, dx. \tag{4.75}$$

Assuming $I_\rho = \rho\delta$ is very small in comparison with both ρ and EI (i.e., we assume δ very small and E very large), the condition that the time derivative of (4.74) with respect to t should be (4.75) leads to (4.26), with $h(x,\xi) = \gamma h(x-\xi)$ there, and the associated boundary conditions. If I_ρ is not neglected we obtain the corresponding Rayleigh model with damping term as in the Euler–Bernoulli model (4.26). Thus the rod model (4.70) is consistent with our earlier developed beam models. Significantly, as we remarked at the end of §4.4, for the damped Rayleigh model the high frequency damping exponent predicted by that model has the same properties as has been observed in the laboratory (Fig. 4.15) and as are predicted from the model equation (4.70).

4.10. Concluding Remarks.

In this paper we have attempted to:

(i) Show that laboratory beams of various types do exhibit modes whose damping exponents are proportional to frequency at the high end of the spectrum;

(ii) Show that there is a relatively simple mathematical model which exhibits comparable spectral behavior and whose form is explainable in terms of plausible physical properties of the beam;

(iii) Show that numerical calculations using a spline based approximation of our mathematical model yield calculated spectral values qualitatively consistent with those obtained from laboratory experimentation and predicted from analysis of the mathematical model.

Some readers may object that the motivation of the damping term in our equation (4.26) is based on a heuristic, plausibility argument rather than on the accepted modus operandi of introducing an appropriate constitutive equation and deriving the dynamics from that point of departure. We refer the reader to [2], [5], [9], [15], [16] for discussions along these lines. All we can say in defense of ourselves on this point is that we have presented sufficient motivation to allow others who specialize in continuum mechanics to carry out such a program if they are inclined to do so. Along the same lines, we also realize that our interpretation of the action of the damping term in terms of fiber structure of the beam, or comparable action of extended crystal structures, may well turn out to be inadequate in the light of other information about the manner in which damping forces act in elastic structures.

We do, without apology, insist that laboratory measurement of damping rates, though necessarily flawed to some extent by exogenous influences, which we have taken some pains to analyze here, must be the final arbiter in deciding between proposed damping mechanisms. Even spotlessly correct theoretical reasoning cannot ultimately take precedence—though it could indicate the need for more careful laboratory work. Hundreds of experiments on our part do indicate frequency proportional damping at the high end of the spectrum and some sort of quadratic behavior, or at least the dominance of a different proportionality constant, at lower frequencies. We invite other interested researchers to carry out similar experimentation and model development complementing, or perhaps even replacing, that reported here.

The reality of near quadratic dependence of the damping exponent on frequency at the low end of the spectrum seems to us to be a very important question. Just citing the models introduced in this chapter, whether or not this phenomenon is real appears to be the decisive factor in choosing the model which we have proposed in preference to "square root" damping as in (4.12) which we have, for the present at least, rejected as not adequately explainable in physical terms—though we have conceded that it could be justified in terms of distances covered by travelling waves of a given frequency. If it could be conclusively demonstrated that low frequency modes are damped at rates which bear the same relationship to frequency as the damping rates associated with higher frequency modes then the whole modeling process would have to be reconsidered in that light. To settle this matter it would seem to be necessary to conduct experiments with very long beams having a large number of low frequency modes.

It also seems to us that the frequency independent damping of higher modes apparently applying to longitudinal rod vibrations is a critical test of the theory. It is particularly significant that this mode of damping is consistent with the very different sort of damping observed in the beam and it is intriguing that the mode of damping which apparently applies for the rod is what is necessary to preserve controllability by means of finite-dimensional controls. If frequency proportional damping exponents did apply to longitudinal vibrations of the rod, those vibrations would be uncontrollable in either the sense standard for the one-dimensional wave equation [20] or that which applies to parabolic processes [8]. The whole area is a most intriguing one. If this chapter should serve to encourage more researchers to enter the area, even if only to refute our contributions, it will have served its purpose.

Note Added in Proof.

Due to circumstances beyond the author's control a period of approximately five years elapsed between the writing of this article and its final publication. During that time other developments have occurred; some reinforcing and others modifying conclusions reached here. For some of those developments we refer the interested reader to [24] and [25], where additional references may also be found.

Acknowledgment.

This research was supported in part by the U. S. Air Force Office of Scientific Research under grants 84-0088 and 85-0283.

References.

[1] S. S. Antman and T.-P. Liu, *Travelling waves in hyperelastic rods*, Quart. Appl. Math., January 1979.

[2] S. S. Antman and R. Malek-Madani, *Dissipative mechanisms*, Proc. of the IMS, Minneapolis, to appear.

[3] C. W. Bert, *Material damping: an introductory review of mathematical models, measures and experimental techniques*, J. Sound and Vibration, 29 (1973), pp. 129–153.
[4] C. W. Bert and R. R. Clary, *Evaluation of experimental methods for determining dynamic stiffness and damping of composite materials*, Special Technical Publications 546, American Society for Testing and Materials, Philadelphia, PA, 1974.
[5] R. E. D. Bishop, *The treatment of damping forces in vibration theory*, J. Royal Aeronautical Society, 59 (1955), pp. 738–742.
[6] G. Chen and D. L. Russell, *A mathematical model for linear elastic systems with structural damping*, Quart. Appl. Math., 39 (1982), pp. 433–454.
[7] E. A. Coddington and N. Levinson, *Theory of Ordinary Differential Equations*, McGraw-Hill, New York, 1955.
[8] H. O. Fattorini and D. L. Russell, *Exact controllability theorems for linear parabolic equations in one space dimension*, Arch. Rational Mech. Anal., 43 (1971), pp. 272–292.
[9] Y. C. Fung, *Foundations of Solid Mechanics*, Prentice-Hall, Englewood Cliffs, NJ, 1965.
[10] H. Goldstein, *Classical Mechanics*, Addison-Wesley, Reading, MA, 1950.
[11] K. Hannsgen and R. Wheeler, *Time delays and boundary feedback stabilization in one-dimensional viscoelasticity*, in Proc. 1986 Vorau Symposium on Identification and Control of Distributed Parameter Systems, to appear.
[12] K. Hannsgen, Y. Renardy, and R. Wheeler, *Effectiveness and robustness with respect to time delays of boundary and distributed stabilization in one-dimensional viscoelasticity*, SIAM J. Control Optim., 26 (1988), pp. 1200–1234.
[13] W. C. Hurty and M. F. Rubinstein, *Dynamics of Structures*, Prentice-Hall, Englewood Cliffs, NJ, 1964.
[14] T. Kato, *Perturbation Theory for Linear Operators*, Springer-Verlag, New York, 1966.
[15] A. L. Kimball and D. E. Lovell, *Internal friction in solids*, Phys. Rev. Ser. 2, 30 (1927), pp. 948–959.
[16] B. J. Lazan, *Damping of Materials and Members in Structural Mechanics*, Pergamon Press, Oxford, 1968.
[17] A. E. H. Love, *A Treatise on the Mathematical Theory of Elasticity*, Fourth ed., Cambridge University Press, Cambridge, 1927.
[18] R. S. Phillips, *Dissipative operators and hyperbolic systems of partial differential equations*, Trans. Amer. Math. Soc., 90 (1959), pp. 193–254.
[19] F. Riesz and B. Sz. Nagy, *Functional Analysis*, F. Ungar, New York, 1955.
[20] D. L. Russell, *Controllability and stabilizability theory for linear partial differential equations*, SIAM Rev., 20 (1978), pp. 639–739.
[21] ——, *On the positive square root of the fourth order derivative operator*, Quart. Appl. Math., 46 (1988), pp. 751–773.
[22] S. P. Timoshenko, *Vibration Problems in Engineering*, Second ed., D. Van Nostrand, Princeton, NJ, 1955.
[23] C. M. Zener, *Elasticity and Anelasticity of Metals*, University of Chicago Press, Chicago, IL, 1948.
[24] D. L. Russell, *A comparison of certain elastic dissipation mechanisms via decoupling and projection techniques*, Quart. Appl. Math., 49 (1991), pp. 373–396.
[25] ——, *A general framework for elastic systems with indirect damping*, J. Math. Anal. Appl., 1992, to appear.

Chapter 5

A Synthesis of Time and Frequency Domain Methods for the Control of Infinite-Dimensional Systems: A System Theoretic Approach

Ruth F. Curtain

5.1. Introduction.

In 1988 I was approached to write an article on frequency domain methods for distributed parameter systems, but as the chapter title reveals, this is not what I have written. Although the finite-dimensional systems and control community has in the past had its state-space advocates and its frequency domain advocates, from the recent research literature it is clear that the "twain" have finally met [86], [84], [64], [100]. This is certainly not the case in the area of infinite-dimensional systems and the wide gap between the books [35] and [77] has not really been bridged by the monograph [5]. I believe that this is now possible and this article has been written in an attempt to show how a *system theoretic approach* can lead to an integrated time and frequency domain treatment of control problems for infinite-dimensional linear systems. Indeed, experience with finite-dimensional control theory indicates that this is a necessary prerequisite to be able to cope successfully with robustness and other performance objectives in more realistic problems of control design.

Although the class of linear infinite-dimensional systems is too large to be covered by one meta theory, it is very useful to identify subclasses which have common mathematical and system theoretic properties and to develop a theory for an appropriate subclass. The most successful class is the class of real, proper, rational transfer functions or the class of linear, time-invariant, finite-dimensional systems usually represented by the matrix quadruple (A, B, C, D). In this chapter, the focus is on two classes of linear infinite-dimensional systems, both with *finite rank inputs and outputs*; the Callier–Desoer class has a frequency domain description, while the Pritchard–Salamon has a state-space description. These two special classes serve to illustrate the advantages of a synthesis of time and frequency domain methods and to suggest ways of extending this approach to more general classes of infinite-dimensional systems.

The well-known *Callier–Desoer* class [28], [29] is essentially the class of transfer functions which can be represented as the sum of an unstable finite-dimensional part and a stable infinite-dimensional part with a smooth impulse response, h, satisfying $\int_0^\infty e^{\alpha t}|h(t)|dt < \infty$ for some $\alpha > 0$. It includes many delay and partial differential equation (p.d.e.) systems and the class has very nice algebraic properties which are useful for control synthesis. In particular, this class is closed under parallel, cascade, and feedback connections which means that it is easy to model complex hybrid and multibody systems as a Callier–Desoer system. Although Balas, Jacobson, and Nett first recognized the need to link the Callier–Desoer class to a state-space description in [116], they only allowed for bounded B and C operators, which is too restrictive for the applications. The appropriate class is the *Pritchard–Salamon* class, which was introduced in [130], [131] as a large class for which the algebraic Riccati equations have unique, nonnegative solutions. It has recently been studied in [39] in connection with the equivalence of input-output stability and exponential stability and in [53] in connection with optimal Hankel norm problems. The definition is a little technical, but from a system theoretic point of view it amounts essentially to a triple (C, A, B) and a triple of Hilbert spaces $W \hookrightarrow X \hookrightarrow V$ such that A generates a consistent C_0-semigroup on all three. The input map B is bounded from $\mathbb{R}^m$ to V and it induces a bounded reachability map from the space of inputs $L_2(0, t_1; \mathbb{R}^m)$ to W. Dually, the output map C is bounded from W to $\mathbb{R}^p$ and it induces a bounded observability map on V. This overlapping of the observability and reachability maps is a smoothness condition, which means that the impulse response is always locally L_1. More general classes with distribution impulse responses exist [137], but the Pritchard–Salamon class is still large enough to include unbounded inputs and outputs in many delay and p.d.e. systems, as shown in [130], [131], and [39]. Moreover, since this class is closed under feedback perturbations, all sorts of hybrid systems with components in the Pritchard–Salamon class will also be in the Pritchard–Salamon class. This class is not contained in the Callier–Desoer class, nor is the latter contained in the former. However, there are close connections between the two, and we shall see that for control applications it is important to be able to go from a frequency domain to state-space description and vice-versa in a rigorous way. By restricting attention to these two relatively simple classes it is possible to address more challenging control problems, such as robust stabilization and control subject to additional performance criteria.

Since infinite-dimensional models are invariably approximate, one must design controllers which stabilize not only the nominal plant, but also perturbations of it. In the engineering literature this property is commonly called *robustness.* Although there are many theories (both state-space and frequency domain) for designing stabilizers for infinite-dimensional systems, there are very few that guarantee robustness of stabilization. Let us recall that even for finite-dimensional systems, although state-feedback controllers designed via the Riccati equation (LQ-controllers) are robust with respect to perturbations, when they are combined with observers to produce the so-called LQG-controller, they need not be robustly stable at all [63]. Since then, robustness has become a central issue in finite-dimensional control and many theories for the design of robust stabilizers for finite-dimensional systems have appeared in the literature ([18], [28], [82], [87], [114], [78], [79], [147], [146], [148], [158]). It is this literature and the era of research on *H^∞-control problems* [75] which has strongly influenced the writing of this article. The same control issues that have been addressed for finite-dimensional systems remain valid for infinite-dimensional systems and indeed some are even more relevant; one can never implement state-feedback for infinite-dimensional systems and the models are nearly always approximate ones. Consequently, it is never enough to have a theory for stabilizers by state-feedback; one needs a theory for robust stabilizers by output feedback. Of course, one needs more, but here I restrict myself to considering the problem of robust stabilization with respect to unstructured perturbations for which the natural setting is in the frequency domain. Two different theories are developed; for the Callier–Desoer class, a theory for robust stabilization with respect to additive perturbations and, for the Pritchard–Salamon class, a theory for robust stabilization with respect to perturbations in normalized coprime factorizations. In addition to obtaining elegant, theoretical solutions, numerical algorithms for designing robust controllers with an a priori robustness margin are obtained. All the numerical calculations are easy matrix manipulations that can be performed with PCMATLAB [129]. The results are interpreted in both the time and frequency domain and one obtains exponential stability of the resulting closed-loop system.

In fact, the two robustness problems considered can be formulated as H^∞-optimal control problems, which have received much attention in the finite-dimensional literature [75]. Although it is not realistic to expect such nice explicit solutions for H^∞-optimal control problems in infinite dimensions, it does suggest that a similar approach could prove promising for their solution, too. This widens the scope of control design considerably, since using an H^∞-formulation one can include many other performance objectives, such as tracking, sensitivity minimization and bandwidth specifications [75].

The above ideas form the basis for the first seven sections. The two special classes of infinite-dimensional systems and several examples are discussed in §§5.2 and 5.3, where their important properties and their relationships to each other are clarified. In particular, it is shown that these classes are closed under feedback configurations, which is an essential prerequisite for control synthesis. For the problem of robust stability under additive perturbations considered in §5.5, the Callier–Desoer class is the appropriate one to consider. Combining this theory with L_∞-approximation results, which are discussed in §5.4, provides an extremely useful approach to designing finite-dimensional robust controllers for infinite-dimensional systems with a priori robustness margins. If the Pritchard–Salamon class also has the property that (C, A) is exponentially detectable and (A, B) is exponentially stabilizable, then it is automatically in the Callier–Desoer class, and this particular subclass is appropriate for H^∞-control problems as pointed out in §5.7; this class has just the right properties for a control synthesis along the lines in [146] and [158]. In §5.6 a special H^∞-problem for this class is solved, namely, robust stabilization for factor perturbations of normalized coprime factorizations, which includes unstable perturbations.

The first version of this article was written in September 1988 and since then several important developments have taken place which confirm the belief expressed then that the system theoretic approach can be extended to a wider class of system than the Callier–Desoer and Pritchard–Salamon classes. Rather than try to rewrite the article completely, I have chosen to add some comments on recent extensions of the two robustness theories at the end of the relevant sections, §§5.5 and 5.6, and on H^∞-control at the end of §5.7. A new section, §5.8, discusses various aspects of different types of robustness and in particular the phenomenon of destabilization of controlled systems by arbitrary small delays. In §5.9 we launch some ideas on how the synthesis of time and frequency domain methods might be extended to more general classes of systems with particular reference to robust stabilization and H^∞-control problems. In addition, some attempt to refer to some related state-space and frequency domain approaches is made. However, a survey of either of these fields is outside the scope of this paper. This article poses more problems that it solves; its main contribution is a personal view on the advantages of a system theoretic synthesis of time and frequency domain methods for infinite-dimensional linear systems and ideas on how to proceed. In fact, this personal view is not so original, but stems from the successful tradition of systems theory for finite-dimensional systems [119]. My hope is that it will stimulate new ideas and open new research directions in the challenging field of infinite-dimensional linear system theory. It has already led to the international conference [46] on "State and Frequency Domain Approaches to Infinite-Dimensional Systems."

5.2. The Callier–Desoer Class of Transfer Matrices

In [28]–[30] Callier and Desoer introduced a class of linear infinite-dimensional systems with finite rank inputs and outputs in terms of their impulse response functions or equivalently their transfer matrices. The impulse response has a smooth part plus countably many impulse terms.

DEFINITION 5.2.1. $\mathcal{A}(\mu)$, $\hat{\mathcal{A}}(\mu)$, $\mathcal{A}_-(\mu)$, $\hat{\mathcal{A}}_-(\mu)$.

For $\mu \in \mathbb{R}$, we say that $f \in \mathcal{A}(\mu)$ if

$$f(t) = \begin{cases} 0, & t < 0 \\ f_a(t) + \sum_{i=0}^{\infty} f_i \delta(t - t_i), & t \geq 0, \end{cases} \tag{5.1}$$

where $f_a(t)e^{-\mu t} \in L_1(0,\infty)$, t_i and f_i are real numbers, $0 = t_0 < t_1 < t_2, < \cdots$, δ represents the delta distribution, and $\sum_{i=0}^{\infty} |f_i| e^{-\mu t_i} < \infty$.

We say that $f \in \mathcal{A}_-(\mu)$ if there exists a $\mu_1 < \mu$ for which $f \in \mathcal{A}(\mu_1)$, and so $\mathcal{A}_-(\mu)$ is a subalgebra of $\mathcal{A}(\mu)$.

$\hat{\mathcal{A}}_-(\mu)$ denotes the class of Laplace transforms of $\mathcal{A}_-(\mu)$; they are functions of a complex variable which are analytic and bounded in $\operatorname{Re} s \geq \mu$.

$\mathcal{A}(\mu)$ and $\hat{\mathcal{A}}_-(\mu)$ are commutative convolution Banach algebras with identity.

Intuitively, $\hat{\mathcal{A}}(0)$ would be a logical choice for the class of *stable systems*; it is a proper subset of systems with a bounded and analytic transfer function in $\operatorname{Re} s > 0$ (the H^∞ class). However, although $\hat{\mathcal{A}}(\mu)$ is a commutative domain with identity, it is not a Bezout domain and so not all elements will have coprime factorizations, [148], [107]. For controller synthesis, coprime factorizations play an essential role [146], and this motivated Callier and Desoer [28] to introduce $\hat{\mathcal{B}}(\mu)$, an algebra of fractions of $\hat{\mathcal{A}}_-(\mu)$, which does admit both left and right coprime factorizations.

DEFINITION 5.2.2. *The Callier–Desoer class* $\hat{\mathcal{B}}(\mu)$.

$$\hat{\mathcal{B}}(\mu) = [\hat{\mathcal{A}}(\mu)].[\hat{\mathcal{A}}_-^\infty(\mu)]^{-1} \tag{5.2}$$

where $\hat{\mathcal{A}}_-^\infty(\mu) = \{\hat{f} \in \hat{\mathcal{A}}_-(\mu)$ and $\hat{f}$ is bounded away from zero at ∞ in $\mathbb{C}_\mu^+ = \{s : \operatorname{Re} s \geq \mu\}\}$. $\hat{f}$ is bounded away from zero at ∞ in $\mathbb{C}_\mu^+$ if there exist $\eta > 0$ and $\rho > 0$ such that $|\hat{f}(s)| \geq \eta$ for all $s \in \mathbb{C}_\mu^+$ such that $|s - \mu| \geq \rho$.

Alternatively, $\mathcal{B}(\mu)$ is the algebra of fractions of $\hat{\mathcal{A}}_-(\mu)$ with respect to $\hat{\mathcal{A}}_-^\infty(\mu)$.

Following the convenient notation from [146], we denote by $\mathcal{M}(\hat{\mathcal{B}}(\mu))$ and $\mathcal{M}(\hat{\mathcal{A}}_-(\mu))$ the class of matrix-valued transfer functions whose elements are in $\hat{\mathcal{B}}(\mu)$ and $\hat{\mathcal{A}}_-(\mu)$, respectively. Intuitively, these correspond to our class of transfer matrices and the transfer matrices that are stable with respect to the stability margin μ, respectively. Notice that these transfer functions are only defined in $\operatorname{Re} s > \mu_1$ for some $\mu_1 < \mu$; they may or may not have meromorphic extensions to the whole complex plane.

Properties of the Callier–Desoer class have been studied at length in [28]–[30]. The following theorem reveals that the transfer matrices belonging to $\mathcal{M}(\hat{\mathcal{B}}(\mu))$ have a very simple decomposition into the sum of a stable (with respect to $\operatorname{Re} s = \mu$) infinite-dimensional part and a totally unstable (with respect to $\operatorname{Re} s = \mu$) finite-dimensional part.

THEOREM 5.2.1 ([28, THM. 3.3], [30, LEMMA 2.1]).

(a) $$\hat{\mathcal{B}}(\mu) = [\hat{\mathcal{A}}_-(\mu)].[\mathcal{R}^\infty(\mu)]^{-1}$$

where $\mathcal{R}^\infty(\mu)$ denotes the class of proper rational functions with complex coefficients which are holomorphic in $\mathbb{C}_\mu^+$ and are bounded away from zero at ∞.

(b) *If $G \in \mathcal{M}(\hat{\mathcal{B}}(\mu))$, then $G = G_+ + G_-$ where $G_- \in \mathcal{M}(\hat{\mathcal{A}}_-(\mu))$ and G_+ is a strictly proper $p \times m$ valued rational matrix; G_+ is zero if and only if $G \in \mathcal{M}(\hat{\mathcal{A}}_-(\mu))$. If $G_+ \neq 0$, then it is the sum of the principal parts of the Laurent expansions of G at its poles in $\mathbb{C}_\mu^+$.*

(c) $G \in \mathcal{M}(\hat{\mathcal{B}}(\mu))$ *if and only if* G *has the decomposition* $G = G_+ + G_-$, *where* $G_- \in \mathcal{M}(\hat{\mathcal{A}}_-(\mu))$ *and* G_+ *is a strictly proper* $p \times m$ *valued rational matrix with all its poles in* $\operatorname{Re}\lambda \geq \mu$.

It is well known that transfer matrices in the Callier–Desoer class always have a state-space realization [77] and the above theorem shows that the Callier–Desoer class is a subset of all linear infinite-dimensional systems. In particular, it excludes systems with infinitely many unstable poles and it excludes systems with a distribution kernel, f_a (we always allow countably many impulses at t_i); examples of such systems are some neutral systems, [136], [107], [108] and some mathematical models of undamped flexible systems, [59], [32]. However, the Callier–Desoer class does include many infinite-dimensional systems including delay systems of neutral and retarded type, parabolic and damped hyperbolic p.d.e. systems, as was realized in [116], [8], [47], [50], [41], [39] and elsewhere. The extent of the Callier–Desoer class is now well understood and its relationship with state-space descriptions will be further clarified in the next section, where we shall describe triples (C, A, B) of operators which have a transfer function $G(s) = C(s - A)^{-1}B \in \mathcal{M}(\hat{\mathcal{B}}(0))$. Here we mention a few examples of different types of transfer functions that occur frequently in the applications, some of which are in the Callier–Desoer class and some of which are not. It is intended as an illustrative rather than a comprehensive list.

Example 5.2.1. *A retarded delay system* [166]

$$\dot{x}_1(t) = x_2(t),$$
$$\dot{x}_2(t) = -\frac{\pi}{2}x_2(t-1) + u(t),$$
$$y(t) = x_1(t-a), \qquad 0 \leq a \leq 1,$$

which has the transfer function

$$G(s) = \frac{e^{-as}}{s\left(s + \frac{\pi}{2}e^{-s}\right)} \in \hat{\mathcal{B}}(0).$$

Example 5.2.2. *A feedthrough delay* [88]

$$\dot{x}(t) = -x(t) + u(t),$$
$$y(t) = x(t-1),$$

which has the transfer function

$$G(s) = \frac{e^{-s}}{1+s} \in \hat{\mathcal{A}}_-(0)$$

and the impulse response

$$h(t) = \begin{cases} 0; & 0 \leq t < 1, \\ e^{-(t-1)}; & t \geq 1. \end{cases}$$

Example 5.2.3. *A neutral system* [108]. Consider the neutral delay system

$$\begin{aligned}\dot{x}_1(t) &= -x_1(t) + u(t),\\ \dot{x}_2(t) - \dot{x}_2(t-r) &= -x_1(t) - a x_2(t),\\ y(t) &= x_2(t),\end{aligned}$$

which has the transfer function

$$G(s) = \frac{1}{(s+1)(s(1-e^{-rs})+a)}.$$

It is in H^∞, but not in $\hat{\mathcal{B}}(0)$.

Example 5.2.4. *Heat equation with Neumann boundary control.* Consider the p.d.e.

$$\frac{\partial z}{\partial t} = \frac{\partial^2 z}{\partial x^2}; \qquad z_x(0) = 0, \quad z_x(1) = u(t),$$

$$y(t) = z(x_0, t).$$

It has the transfer function

$$G(s) = \frac{\cosh \sqrt{s} x_0}{\sqrt{s} \sinh \sqrt{s}} \in \hat{\mathcal{B}}(0) \text{ and } \hat{\mathcal{A}}_-(-\epsilon) \quad \text{for } \epsilon > 0.$$

Example 5.2.5. *Heat equation with Dirichlet boundary control* [55]. Consider the p.d.e.

$$\frac{\partial z}{\partial t} = \frac{\partial^2 z}{\partial x^2}; \quad z(0) = 0, \quad z(1) = u(t),$$

$$y(t) = z(x_0, t).$$

It has the transfer function

$$G(s) = \frac{\sinh \sqrt{s} x_0}{\sinh \sqrt{s}} \in \hat{\mathcal{A}}_-(0).$$

Example 5.2.6. *Wave equation* [137]. Consider the p.d.e.

$$\frac{\partial^2 z}{\partial t^2} = \frac{\partial^2 z}{\partial x^2},$$

$$z(0,t) = 0, \qquad \frac{\partial z}{\partial x}(1,t) = u(t),$$

$$y(t) = \frac{\partial z}{\partial t}(1,t).$$

This has the transfer function $G(s) = (1-e^{-2s})/(1+e^{-2s}) \notin \hat{\mathcal{B}}(0)$; the impulse response is a distribution. It also has infinitely many poles on the imaginary axis.

Example 5.2.7. *Undamped flexible beam* [137], [40], [32]. Consider the one-dimensional fourth-order hyperbolic p.d.e. described by

$$\begin{cases} \dfrac{\partial^2 z}{\partial t^2}(x,t) = -a^4 \dfrac{\partial^4 z}{\partial x^4}(x,t); \qquad 0 \le x \le 1 \\ z(0,t) = 0, \quad \dfrac{\partial^2 z}{\partial x^2}(1,t) = 0, \quad \dfrac{\partial z}{\partial x}(0,t) = 0, \quad \dfrac{\partial^3 z}{\partial x^3}(1,t) = bu(t), \\ y_1(t) = c\dfrac{\partial z}{\partial t}(1,t), \qquad y_2(t) = c\dfrac{\partial^2 z}{\partial t\, \partial x}(1,t). \end{cases}$$

Then the transfer functions for the two outputs are [149]:

$$G_1(s) = \frac{sbc}{\lambda^3}\left(\frac{2\sinh\lambda\cos\lambda - 2\cosh\lambda\sin\lambda}{1+\cos\lambda\cosh\lambda}\right), \qquad \lambda^4 = \frac{-s^2}{a^4},$$

$$G_2(s) = \frac{2s(\sinh\lambda\cos\lambda + \cosh\lambda\sin\lambda)}{\lambda(1+\cos\lambda\cosh\lambda)}.$$

Examples 5.2.1, 5.2.2, 5.2.4, and 5.2.5 are in the Callier–Desoer class, $\hat{\mathcal{B}}(0)$; they are the sum of a (possibly zero) finite-dimensional, unstable part, and the stable part has an integrable impulse response $f_a \in \mathcal{A}_-(0)$. We could also add countably many impulses $f_i\delta(t-t_i)$ as in (5.1).

Example 5.2.1 represents the classes of delay systems recently studied in [166], [128]. They have finitely many unstable poles and infinitely many stable poles which asymptote typically like $\alpha \log n + \beta i n$ and so are not contained in any sector [12].

Example 5.2.2 represents the class $e^{-s}G_0(s)$, where G_0 is rational, studied in [88], [89]; these only have finitely many poles.

Examples 5.2.4 and 5.2.5 are parabolic p.d.e. examples which have finitely many poles in any open half plane.

Example 5.2.3 is an example of a neutral system which is in H^∞, but not in $\hat{\mathcal{B}}(0)$. It has a sequence of poles in $\operatorname{Re} s < 0$ which asymptote to the imaginary axis, so G cannot have a decomposition such as in Theorem 5.2.1(c). Other interesting examples of neutral systems are discussed in [108].

Examples 5.2.6 and 5.2.7 are p.d.e. systems that are not in the Callier–Desoer class because the f_a-term in its stable part (cf. (5.1)) is not an L_1-function, but a distribution. (Notice that we still allow countably many impulse distributions in the $f_i\delta(t-t_i)$ terms in (5.1).) Another reason that Examples 5.2.6 and 5.2.7 are excluded from the Callier–Desoer class is that they have infinitely many poles on the imaginary axis; this is often the case for mathematical p.d.e. models (see [59]). $G_2(s)$ of Example 5.2.7 even has the property that $|G(x)| \to \infty$ as $x = \operatorname{Re} s \to \infty$. In fact, both Examples 5.2.6 and 5.2.7 are improper in the sense of the following definition.

DEFINITION 5.2.3. Proper and improper systems

(i) Transfer matrices G in $\mathcal{M}(\hat{\mathcal{B}}(\mu))$ are said to be *proper* if $\lim G(s)$ exists as $|s| \to \infty$ in $\operatorname{Re} s > 0$. We denote this limit by $G(\infty)$.

(ii) If this limit is the zero matrix, then we say G is *strictly proper.*

(iii) If there exists a sequence s_n in $\operatorname{Re} s > 0$ such that $|s_n| \to \infty$ and $|G(s_n)| \to \infty$, then we call G *improper.*

We remark that the requirement that G have a limit at infinity is not satisfied by all elements of $\hat{\mathcal{B}}(\mu)$. For example, e^{-s} is neither proper, nor improper in the sense of the above definition. It is, however, bounded at infinity, as are all elements in $\mathcal{M}\hat{\mathcal{B}}(\mu)$.

A very successful approach to controller synthesis for rational transfer functions is to exploit its algebraic properties and in particular, coprime factorizations over the class of stable, proper rational transfer functions. This idea was conceived in [61] and in [146] Vidyasagar gives a self-contained introduction to this approach for rational transfer functions and applies it to various control problems including robust control (see §§5.5 and 5.6) and H^∞-control (see §5.7). The same ideas can be applied to any ring that has coprime factorizations and this was the underlying motivation to introduce the Callier–Desoer class [28], [29]. For this class the following definitions of coprime factorizations are appropriate.

DEFINITION 5.2.4. If $G \in \mathcal{M}(\hat{\mathcal{B}}(\mu))$, then $\tilde{M}^{-1}\tilde{N}$ is called a *left coprime factorization* of G if and only if

(i) $\tilde{M}$ and $\tilde{N} \in \mathcal{M}(\hat{\mathcal{A}}_-(\mu))$ and $\det \tilde{M} \in \hat{\mathcal{A}}_-^\infty(\mu)$;

(ii) $G = \tilde{M}^{-1}\tilde{N}$;

(iii) $\tilde{M}$ and $\tilde{N}$ are left coprime in the sense that there exist X and Y in $\mathcal{M}(\hat{\mathcal{A}}_-(\mu))$ such that

$$\tilde{N}Y - \tilde{M}X = I. \tag{5.3}$$

$\tilde{M}^{-1}\tilde{N}$ is called a *normalized* left coprime factorization of G if it is a left coprime factorization and

$$\tilde{N}(j\omega)\tilde{N}^t(-j\omega) + \tilde{M}(j\omega)\tilde{M}^t(-j\omega) = I \quad \text{for all real } \omega. \tag{5.4}$$

The definition of a right coprime factorization is analogous. Theorem 5.2.1(b) reveals a simple way of finding coprime factorizations of G based on those of the finite-dimensional part: if $G_+ = \tilde{M}_0^{-1}\tilde{N}_0$ represents a left coprime factorization of the finite-dimensional part, G_+, then $G = \tilde{M}_0^{-1}(M_0G_- + \tilde{N}_0)$ is a left coprime factorization of G, where G_- is the stable infinite-dimensional part of G.

The Callier–Desoer class has just the right properties needed to apply the algebraic control system synthesis exposed in [61], [146], and [158] (see §8.3). The basic problem is to design a feedback controller, K, to stabilize the plant, G, as in Fig. 5.1, by which we mean the following.

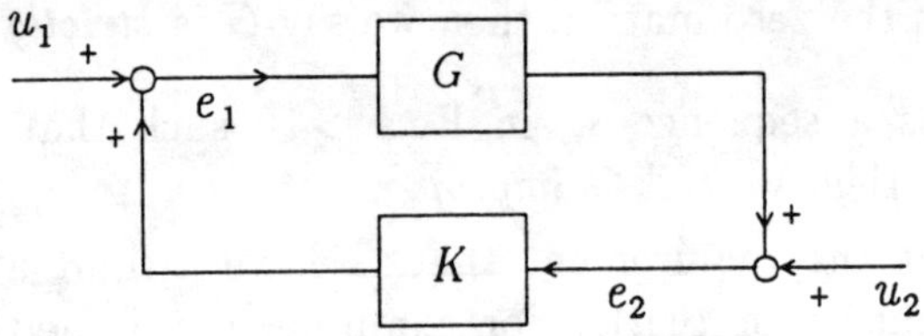

FIG. 5.1. *Internal stability.*

DEFINITION 5.2.5. The feedback system (G, K) of Fig. 5.1 with $G, K \in \mathcal{M}(\hat{\mathcal{B}}(0))$ is said to be *input-output stable* if and only if (a) and (b) are satisfied:

(a) $S = (I - GK)^{-1}$, KS, SG, $I - KSG \in \mathcal{M}(\hat{\mathcal{A}}_-(0))$;

(b) $\inf_{\mathbb{C}^+} |\det(I - GK)(s)| > 0$.

Notice that our definition of stability implies that the closed-loop system has no singularities in $\operatorname{Re} s \geq 0$. Furthermore, we have assumed that both G and K belong to the Callier–Desoer class, $\mathcal{M}(\hat{\mathcal{B}}(0))$, and so they both possess left and right coprime factorizations. This makes it possible to parametrize all stabilizing controllers just as in the finite-dimensional case.

THEOREM 5.2.2 ([146, THM. 8.3.12]). *Suppose that $G \in \mathcal{M}(\hat{\mathcal{B}}(0))$ has the right and left coprime factorizations $G = NM^{-1} = \tilde{M}^{-1}\tilde{N}$ such that X, Y, $\tilde{X}$, $\tilde{Y} \in \mathcal{M}(\hat{\mathcal{A}}_-(0))$ satisfy*

$$\tilde{N}Y - \tilde{M}X = I; \quad -\tilde{Y}N + \tilde{X}M = I \quad \text{in } \operatorname{Re} s \geq 0. \tag{5.5}$$

Then all stabilizing controllers $K \in \mathcal{M}(\hat{\mathcal{B}}(0))$ are given by

$$\begin{aligned} K &= (Y + MJ)(X + NJ)^{-1} \\ &= (\tilde{X} + J\tilde{N})^{-1}(\tilde{Y} + JM) \end{aligned} \tag{5.6}$$

whenever these expressions are well defined and $J \in \mathcal{M}(\hat{\mathcal{A}}_-(0))$.

We shall see in §§5.6 and 5.7 that this parametrization plays an important role in the formulation of H^∞-control problems, as does the concept of doubly coprime factorizations.

DEFINITION 5.2.6 (DOUBLY COPRIME FACTORIZATION). We say the $G \in \mathcal{M}(\hat{\mathcal{B}}(0))$ has a *doubly coprime factorization* if there exist N, M, $\tilde{N}$, $\tilde{M}$, X, $\tilde{X}$, Y, $\tilde{Y} \in \mathcal{M}(\hat{\mathcal{A}}_-(0))$ such that

$$G = NM^{-1} = \tilde{M}^{-1}\tilde{N}, \tag{5.7}$$

$$\begin{bmatrix} \tilde{X} & -\tilde{Y} \\ -\tilde{N} & \tilde{M} \end{bmatrix} \begin{bmatrix} M & Y \\ N & X \end{bmatrix} = \begin{bmatrix} I & 0 \\ 0 & I \end{bmatrix}. \tag{5.8}$$

In the next section we shall give explicit formulas for doubly coprime factorizations for systems which have a smooth state-space realization $C(sI - A)^{-1}B$ for the transfer function.

Finally, some comments about our choice of the Callier–Desoer class of transfer functions are in order. We saw that it does not include some interesting infinite-dimensional systems such as Examples 5.2.6 and 5.2.7 due to the fact that it is restricted to those systems with at most finitely many unstable poles. While this may seem a restriction on our choice for our class of transfer functions, the following result indicates that, from a practical design point of view, it is not such a restriction.

THEOREM 5.2.3. *Suppose that G and $K \in \mathcal{M}(\hat{\mathcal{A}}_-(\mu))$ for some real μ. If K stabilizes G in the sense of Definition* 5.2.5 *and GK is strictly proper in $\overline{\mathbb{C}^+_\mu}$, then $G \in \mathcal{M}(\hat{\mathcal{B}}(0))$.*

Proof. See Theorem 6.1 in [91].

In practice, due to physical limitations on sensors and actuators, one can only implement strictly proper controllers. So, in practice, the assumptions of Theorem 5.2.3 will be satisfied by many examples with infinitely many unstable poles, including Example 5.2.6. Theorem 5.2.3 tells us that it is not possible to stabilize this system by means of a strictly proper controller from $\mathcal{M}(\hat{\mathcal{A}}_-(\mu))$. However, Example 5.2.8 is not in $\mathcal{M}(\hat{\mathcal{A}}_-(\mu))$ for any μ and this cautions us that not all systems are covered by Theorem 5.2.3.

5.3. The Pritchard–Salamon Class of State-Space Systems

For control synthesis it is often convenient to have an abstract state-space formulation in terms of operators (C, A, B). Of the many competing formulations, the one which suits our purposes is that introduced by Salamon in [136] for neutral systems and extended to partial differential systems in [137]. (See also [150]–[152] and [43] for various extensions.) This class is larger than the Callier–Desoer class, and so we restrict ourselves to a subclass of "smooth" systems with finite rank inputs and outputs which were studied in [130], [131] in connection with the linear quadratic control problem and in [39] in connection with the equivalence of internal and external stability. This class of systems is now well understood (see [111] and [52]). We do not attempt to discuss all the properties in detail here. In particular, they always have an integrable impulse response. However, they will not belong to the Callier–Desoer class $\mathcal{M}(\hat{\mathcal{B}}(0))$ if A has infinitely many unstable eigenvalues and so neither class is contained in the other. The precise relationships will be explained in this section.

DEFINITION 5.3.1 (THE PRITCHARD–SALAMON CLASS). Suppose there exist three separable Hilbert spaces V, X, and W with continuous, dense injections and that A generates a C_0-semigroup, $S(t)$, on V, X and W which is consistent in the sense that the restriction of $S(t)$ to W (or X) defines a C_0-semigroup on W (or X) and so we use the one symbol for both $S(t)$ and its generator, A. We further assume that

$$Z = D_V(A) \hookrightarrow W \hookrightarrow X \hookrightarrow V \tag{5.9}$$

where $D_V(A)$ denotes the domain of A considered as an operator on V.

We suppose that $B \in \mathcal{L}(\mathbb{R}^m, V)$ induces a smooth controllability map with respect to W on $[0, t_1]$ for all finite t_1

$$\left\| \int_0^{t_1} S(t_1 - s)Bu(s)ds \right\|_W \leq \beta \|u\|_{L^2(0,t_1;\mathbb{R}^m)} \tag{5.10}$$

and that $C \in \mathcal{L}(W, \mathbb{R}^p)$ induces a smooth observability map with respect to V on $[0, t_1]$ for all finite t_1,

$$\|(CS(\cdot)x\|_{L_2(0,t_1;\mathbb{R}^p)} \le \gamma \|x\|_V. \tag{5.11}$$

If (C, A, B) satisfies the above assumptions, we say that it belongs to the *Pritchard–Salamon* class.

We remark that the role of X is to some extent artificial and its only purpose is to define a dual system (B^*, A^*, C^*). To do this we identify the duals of U, Y, and X. Using X as a pivot space, we take V^* to be the space of all linear functionals on V which are continuous with respect to the topology on X. W^* is defined analogously.

Consequently (5.10) and (5.11) are dual conditions in the sense that B satisfies a (5.10) condition with respect to S if and only if B^* satisfies a (5.11) condition with respect to S^*. It is a technical assumption which is satisfied in all known examples (see [129] for further discussions on this point), and, moreover, it has the important consequence that $S(t)$ has the same exponential growth rate on V, X, and W.

We remark that Examples 5.2.1 and 5.2.4 belong to the Pritchard–Salamon class, but Examples 5.2.5–5.2.7 do not. The following p.d.e. beam model has been used in various robustness studies in [23]–[26]. To show that it belongs to the Pritchard–Salamon class, we use a spectral analysis which was developed in [39], which is an extension of a similar analysis used in [93] and [130] for parabolic and undamped hyperbolic systems.

Example 5.3.1. *Euler–Bernoulli beam with Kelvin–Voigt damping* [20], [21]. Consider the following p.d.e. describing the deflection $w(x, t)$ at position x at time t.

$$\begin{aligned}
&w_{tt} + \alpha_1 A w_t + \alpha_2 A w = \frac{1}{\rho a}(\delta(0)u_1(t) - \delta(0)u_2(t)),\\
&w_{xx}(-1) = 0 = w_{xx}(1); \qquad w_{xxx}(-1) = 0 = w_{xxx}(1),\\
&-1 \le x \le 1; \qquad t \ge 0,\\
&y_1(t) = w(0,t); \qquad y_2(t) = w_x(0,t),
\end{aligned}$$

where α_1 and α_2 are the damping and stiffness coefficients, ρa the mass per unit length, and A is the positive, selfadjoint operator given by

$$A = \frac{d^4}{dx^4},$$

$$D(A) = \left\{ \begin{array}{c} h \in L_2(-1,1) : w_x, w_{xx}, w_{xxx}, w_{xxxx} \in L_2(-1,1) \\ w_{xx}(-1) = w_{xx}(1) = 0, w_{xxx}(-1) = 0 = w_{xxx}(1) \end{array} \right\}.$$

As state space we choose $X = D(A^{1/2}) \oplus L_2(-1,1)$ with the inner product

$$\left\langle \begin{pmatrix} v_1 \\ v_2 \end{pmatrix}, \begin{pmatrix} w_1 \\ w_2 \end{pmatrix} \right\rangle_H = \langle v_1, w_1 \rangle_{L_2} + \langle A^{1/2} v_1, A^{1/2} w_1 \rangle_{L_2} + \langle v_2, w_w \rangle_{L_2}$$

so that on X this p.d.e. can be formulated as the abstract system

$$\dot{z} = \mathcal{A}z + Bu; \qquad y = Cz$$

where

$$\mathcal{A} = \begin{pmatrix} 0 & I \\ -\alpha_2 A & -\alpha_1 A \end{pmatrix}; \qquad B = \frac{1}{\rho a}\begin{pmatrix} 0 & 0 \\ \delta(0) & -\delta'(0) \end{pmatrix},$$

$$C\begin{pmatrix} v_1 \\ v_2 \end{pmatrix} = \begin{pmatrix} v_1(0) \\ \frac{dv_1}{dx}(0) \end{pmatrix}; \qquad z = \begin{pmatrix} w \\ w_t \end{pmatrix}.$$

A convenient way to verify the conditions (5.10) and (5.11) is to do a spectral analysis based on the eigenvectors of the system operator $\mathcal{A}$ (see [36], [21]). $\mathcal{A}$ has the eigenvalues μ_n, μ_{-n}, $n = 1, 2 \cdots$, given by

$$\mu_n, \mu_{-n} = \frac{-\alpha\lambda_n^4 \pm \sqrt{\alpha^2\lambda_n^8 - 4\lambda_n^4}}{2}$$

where λ_n are the roots of

$$\begin{aligned} \sinh\lambda_n \cos\lambda_n + \cosh\lambda_n \sin\lambda_n = 0; \qquad n = 1, 3, 5, 7, \\ \sinh\lambda_n \cos\lambda_n - \cosh\lambda_n \sin\lambda_n = 0; \qquad n = 2, 4, 6, 8. \end{aligned}$$

Notice that $\lambda_1 = 0 = \lambda_2$, which implies that $\mathcal{A}$ has 4 eigenvalues at the origin; the rest are in $\operatorname{Re} s < 0$.

The normalized eigenvectors are

$$\varphi_n = \eta_n \begin{bmatrix} v_n \\ \mu_n v_n \end{bmatrix}, \qquad \varphi_{-n} = \eta_{-n} \begin{bmatrix} v_n \\ \mu_{-n} v_n \end{bmatrix} \quad \text{with } \eta_{\pm n} = (1+\lambda_n^4+|\mu_{\pm n}|^2)^{-\frac{1}{2}}$$

where for even n

$$v_n = \alpha_n \left\{ \cos\lambda_n + \cos\lambda_n \frac{\cosh\lambda_n x - \cos\lambda_n x}{\cosh\lambda_n + \cos\lambda_n} \right\},$$

$$\alpha_n = \frac{\cosh\lambda_n + \cos\lambda_n}{\sqrt{\cosh^2\lambda_n + \cos\lambda_n^2}}$$

and there are similar expressions for odd n. $\{\phi_n, \phi_{-n})$ forms a Riesz basis for X.

The eigenvectors of $\mathcal{A}^*$ are

$$f_n = \begin{pmatrix} v_n \\ -\left(\frac{1+\lambda_n^4}{\lambda_n^4}\right)\bar{\mu}_n v_n \end{pmatrix}; \qquad f_{-n} = \begin{pmatrix} v_n \\ -\left(\frac{1+\lambda_n^4}{\lambda_n^4}\right)\bar{\mu}_{-n} v_n \end{pmatrix}$$

and $\{\varphi_n, \psi_n\}$ form a biorthogonal sequence in X where

$$\psi_n = \frac{1}{\langle f_n, \varphi_n \rangle_H} f_n.$$

We need to know that B^* is given by

$$B^* \begin{pmatrix} w_1 \\ w_2 \end{pmatrix} = \begin{pmatrix} w_2(0) \\ \dfrac{d}{dx} w_2(0) \end{pmatrix}.$$

In [39] it is shown that there exist Hilbert spaces V and W such that (C, A, B) is a Pritchard–Salamon system if the following holds

$$\sum_{-\infty}^{\infty} \frac{c_n b_n}{|\mathrm{Re}\, \mu_n|^{\frac{1}{2}}} < \infty$$

where $c_n = \|C\varphi_n\|$; $b_n = \|B^*\psi_n\|$.

Using asymptotic estimates it is readily verified that this indeed holds and that (C, A, B) is a Pritchard–Salamon system with respect to $V = H_\beta$ and $W = H_\gamma$, where H_β and H_γ are scaled spaces of the type

$$H_\alpha = \left\{ x = \sum_{-\infty}^{\infty} x_n \phi_n : \sum_{-\infty}^{\infty} \alpha_n |x_n|^2 < \infty \right\}$$

for the positive sequence $\{\alpha_n\}$; $n = \pm 1, \cdots, \infty$, with inner product

$$\langle x, h \rangle_\alpha = \sum_{-\infty}^{\infty} \alpha_n x_n h_n; \qquad x, h \in H_\alpha.$$

The sequences $\{\beta_n\}$ and $\{\gamma_n\}$ for this example are given by

$$\gamma_n = \frac{|\mathrm{Re}\, \mu_n|^{\frac{1}{2}} c_n}{b_n}, \qquad \beta_n = \frac{c_n}{b_n |\mathrm{Re}\, \mu_n|^{\frac{1}{2}}}.$$

As a corollary of this spectral analysis we obtain the following expansions for the transfer function, G and the impulse response, h,

$$G(s) = \sum_{-\infty}^{\infty} \frac{C\varphi_n \overline{B^*\psi_n}}{s - \mu_n},$$

$$h(t) = \sum_{-\infty}^{\infty} e^{\mu_n t} C\phi_n \overline{B^*\psi_n} + t\, C\varphi_1 \overline{B^*\psi_{-1}} + t\, C\varphi_2 \overline{B^*\psi_{-2}},$$

and using asymptotic estimates on this last expression for h shows that the system also belongs to the Callier–Desoer class (Definition 5.2.1). Notice that, strictly speaking, the above expansion for G only holds where the resolvent of $\mathcal{A}$ is analytic, in $\mathrm{Re}\, s > 0$.

One can sometimes obtain more information about the transfer function using a frequency domain approach [13]. For this example, the following closed-form expression was obtained in [149]

$$G(s) = \frac{\lambda(1 + \cos\lambda \cosh\lambda)}{s^2(\cosh 2\lambda \cos 2\lambda - 1)} \begin{bmatrix} r(\lambda) & 0 \\ 0 & \lambda^2(\cos\lambda \sinh\lambda + \sin\lambda \cosh\lambda) \end{bmatrix}$$

where $r(\lambda) = \cos\lambda \sinh\lambda - \sin\lambda \cosh\lambda$ and $\lambda^4 = -s^2/(1+s\alpha)$. G is in fact meromorphic in the punctured complex plane $\mathbb{C}\backslash\{-1/\alpha\}$ with an essential singularity at $-1/\alpha$. This closed-form expression also gives information about the pole-zero pattern which is important for control applications.

Of course, it is well known that there are other ways of establishing a well-posed state-space formulation, and these are discussed elsewhere in this volume. (See also [137], [150]–[152] and the references therein.) However, for our synthesis we use conditions (5.9)–(5.11) repeatedly in our analysis and the spectral analysis approach illustrated in Example 5.3.1 is a convenient way of verifying these conditions for simple examples in the Pritchard–Salamon class. As a corollary you obtain a series expansion for the transfer functions and the impulse response. This spectral analysis is a straightforward application of techniques discussed in earlier papers [93], [21], [130], [36], [39], and [93]. Methods for verifying (5.9)–(5.11) for delay systems are different, and we refer to [131] for an account of this.

The most useful property of Pritchard–Salamon systems is that they are closed under feedbacks.

LEMMA 5.3.1 ([136], [137]). (a) *For $F \in \mathcal{L}(V, \mathbb{R}^m)$, if assumption* (5.10) *holds, $A + BF$ generates a C_0-semigroup $S_F(t)$ on W and V and B satisfies a* (5.10) *condition with respect to $S_F(t)$. If C satisfies* (5.11) *with respect to $S(t)$, then it will satisfy a* (5.11) *condition with respect to $S_F(t)$ too.*

(b) *For $H \in \mathcal{L}(\mathbb{R}^p, W)$ if* (5.11) *holds, $A + HC$ generates a C_0-semigroup $S_H(t)$ on W and V and C satisfies a* (5.11) *condition with respect to $S_H(t)$. If B satisfies* (5.10) *with respect to $S(t)$, then it will satisfy a* (5.10) *condition with respect to $S_H(t)$ too.*

More subtle perturbation results are contained in [52]. We remark that "$A + BF$ generates a C_0-semigroup on X" should be interpreted in the sense that the following integral equation has a unique solution $S_F(t)x$ for all $x \in X$

$$S_F(t)x = S(t)x + \int_0^t S(t-\beta)BFS_F(\beta)x\,d\beta \tag{5.12}$$

and $S_F(t)$ is a C_0-semigroup on X.

Notice that in (a) of Lemma 5.3.1 $D_V(A+BF) = D_V(A)$, since $BF \in \mathcal{L}(V)$ and so $S_F(t)$ is defined on W, X and V and has the same decay rate on all three spaces.

The conditions (5.10) and (5.11) allow us to consider the state-space system defined by

$$\dot{x} = Ax + Bu; \qquad x(0) = x_0, \tag{5.13i}$$

$$y = Cx. \tag{5.13ii}$$

This is well posed in the sense that for $u \in L_2(0, t_1; \mathbb{R}^m)$, (5.13i) has the unique mild solution $x \in C(0, t_1; X)$ for all $x_0 \in X$ given by

$$x(t) = S(t)x_0 + \int_{t_0}^t S(t-\beta)Bu(\beta)d\beta \tag{5.14}$$

and so in (5.13ii) $y \in C(0, t_1; \mathbb{R}^p)$ [136].

Condition (5.9) is used in solving the linear quadratic control problem [130].

In [37], [39] it is shown that if $S(t)$ is exponentially stable, then the reachability map can be extended to a bounded map $\mathcal{B} : L_2(0, \infty; \mathbb{R}^m) \to W$ given by

$$\mathcal{B}u := \int_0^\infty S(t)Bu(t)dt. \tag{5.15}$$

Similarly, the observability map can be extended to a bounded map $\mathcal{C} : V \to L_2(0, \infty; \mathbb{R}^p)$ given by

$$\mathcal{C}x := CS(t)x. \tag{5.16}$$

This also implies that the *Hankel operator* Γ induced by an exponentially stable Pritchard–Salamon system (C, A, B) is a bounded operator from $L_2(0, \infty; \mathbb{R}^m)$ to $L_2(0, \infty; \mathbb{R}^p)$ which is given by

$$\begin{aligned}(\Gamma u)(t) &= (\mathcal{C}\mathcal{B}u)(t) \\ &= CS(t)\int_0^\infty S(\beta)Bu(\beta)d\beta \\ &= \int_0^\infty CS(t+\beta)Bu(\beta)d\beta \qquad \text{(see [52]).}\end{aligned} \tag{5.17}$$

In [37] it is proved that Pritchard–Salamon systems have a Hilbert–Schmidt Hankel operator. So Γ is compact and it has countably many singular values, $\{\sigma_i\}$ which are equal to the square roots of the eigenvalues of the positive, selfadjoint operator $\Gamma^*\Gamma$ and are called the *Hankel singular values* of the system (C, A, B), and they are important system theoretic invariants.

Relationships between the Callier–Desoer class of transfer functions and state-space representations were first discussed in [116] for systems (C, A, B), where A generates a C_0-semigroup on a Hilbert space X and B and C are finite-rank, *bounded* maps on X. This class is too small for interesting applications, where the input and output maps are usually unbounded. The appropriate class is the Pritchard–Salamon class whose relationship with the Callier–Desoer class was clarified in [39], [111], and [52] by the following result.

LEMMA 5.3.2. (a) *If* (A, B) *satisfies* (5.10), *then* $(s - A^v)^{-1}B \in \mathcal{L}(\mathbb{R}^m, W)$ *for* $\text{Re}\, s > \alpha$ *and* $C(s - A^v)^{-1}B \in \mathcal{L}(\mathbb{R}^m, \mathbb{R}^p)$ *for* $C \in \mathcal{L}(W, \mathbb{R}^p)$ *and it is analytic for* $\text{Re}\, s > \alpha$ *(*α *is the maximum of the growth constants of* $S(t)$ *on* W *and on* V*). Moreover,*

$$\int_0^\infty e^{-\beta t}|CS(t)B|dt < \infty \quad \textit{for } \beta > \alpha. \tag{5.18}$$

(b) *Similarly, if* (C, A) *satisfies* (5.3), $C(s - A^w)^{-1} \in \mathcal{L}(V, \mathbb{R}^p)$ *for* $\text{Re}\, s > \alpha$ *and* $C(s - A^w)^{-1}B \in \mathcal{L}(\mathbb{R}^m, \mathbb{R}^p)$ *for all* $B \in \mathcal{L}(\mathbb{R}^m, V)$ *and it is analytic for* $\text{Re}\, s > \alpha$ *and* (5.18) *holds for* $\beta > \alpha$.

(c) $C(s - A^w)^{-1}B = C(s - A^v)^{-1}B$ *for* $\text{Re}\, s > \alpha$.

Part (c) shows that the transfer function is unambiguously defined and so we usually write $C(s - A)^{-1}B$. Inequality (5.18) shows that the impulse response is always integrable and so $C(s-A)^{-1}B \in \mathcal{M}(\hat{\mathcal{A}}_-(\mu))$ for $\mu > \alpha$, but it is more important to know if $C(s-A)^{-1}B \in \mathcal{M}(\hat{\mathcal{B}}_-(0))$. Clearly it is sufficient if A has finitely many unstable eigenvalues, but it is also interesting to place this in a system theoretic context.

LEMMA 5.3.3 ([39, THM. 4.1]). *Let (C, A, B) belong to the Pritchard–Salamon class. Then $C(s-A)^{-1}B \in \mathcal{M}(\hat{\mathcal{B}}_-(0))$ if either (A, B) is exponentially stabilizable by a feedback $F \in \mathcal{L}(V, \mathbb{R}^m)$ or (C, A) is exponentially detectable by an output injection $H \in \mathcal{L}(\mathbb{R}^p, W)$.*

In fact, the classes of exponentially stabilizable (or detectable) Pritchard–Salamon systems have properties reminiscent of finite-dimensional systems, as is clarified by the following result.

LEMMA 5.3.4 ([118], [98], [39]). *If (A, B) belongs to the Pritchard–Salamon class, then (A, B) is exponentially stabilizable by $F \in \mathcal{L}(V, \mathbb{R}^m)$ if and only if A has finitely many eigenvalues of finite multiplicity in $\operatorname{Re} s \geq 0$ and with respect to the spectral decomposition $(A, B) = \left(\begin{pmatrix} A_1 & 0 \\ 0 & A_2 \end{pmatrix}, \begin{bmatrix} B_1 \\ B_2 \end{bmatrix}\right)$, where A_1 contains all the unstable eigenvalues of A, the following holds*

(i) *(A_1, B_1) is controllable,*

(ii) *A_2 generates an exponentially stable semigroup on V.*

This type of result was first recognized for the case of bounded inputs B in [118] and [98] and for a similar class of unbounded inputs in [60]; it is a consequence of the finite-rank input. Sufficient conditions for A_2 to generate an exponentially stable semigroup on W and V are given in [52]. Dual results hold for exponential detectability. In [52], they also consider stabilizability by weak feedbacks $F \in \mathcal{L}(W, U)$, but here we only consider feedbacks $F \in \mathcal{L}(V, Z)$.

More important still is the relationship between frequency domain stability and the exponential stability of A. Frequency domain stabilization techniques lead to a transfer function in H^∞, which need not imply that the closed-loop system has a system operator which generates an exponentially stable C_0-semigroup. Sufficient conditions for this were proved in [116] and [98] for the case of bounded B and C and were extended to the unbounded case in [39] and [52].

THEOREM 5.3.1 ([39], [52]). *Suppose that (C, A, B) is a Pritchard–Salamon system, which is exponentially stabilizable on V and detectable on W. Then A is exponentially stable on V and W if and only if $G(s) = C(s-A)^{-1}B \in \mathcal{M}(\hat{\mathcal{A}}_-(0))$.*

So we have seen that the Callier–Desoer class and the Pritchard–Salamon class are closely related via the concepts of stabilizability and detectability. That these concepts are also related to the existence of doubly coprime factorizations was already recognized in [103] in a ring setting and [117] for the case of state-space infinite-dimensional systems with bounded B and C. Their result was extended to the Pritchard–Salamon class in [42].

LEMMA 5.3.5 ([42, LEMMA 4.10]). *Suppose (C, A, B) belongs to the Pritchard–Salamon class and (C, A) is exponentially detectable on W and*

(A, B) is exponentially stabilizable on V. Then $G(s) = D + C(s - A)^{-1}B$ has the following doubly coprime factorization of the form (5.7), (5.8), *where*

$$(5.19)\quad \begin{aligned} M(s) &= I + F(s - A_F)^{-1}B; & \tilde{M}(s) &= I + C(s - A_H)^{-1}H,\\ N(s) &= D + C_F(s - A_F)^{-1}B; & \tilde{N}(s) &= D + C(s - A_H)^{-1}B_H,\\ X(s) &= I - C_F(s - A_F)^{-1}H; & \tilde{X}(s) &= I - F(s - A_H)^{-1}B_H,\\ Y(s) &= -F(s - A_F)^{-1}H; & \tilde{Y}(s) &= -F(s - A_H)^{-1}H. \end{aligned}$$

M, N, $\tilde{M}$, $\tilde{N}$, X, Y, $\tilde{X}$, $\tilde{Y} \in \mathcal{M}(\hat{\mathcal{A}}_-(0))$ and $F \in \mathcal{L}(V, \mathbb{R}^m)$ is chosen such that $A_F = A + BF$ generates an exponentially stable semigroup on V and $H \in \mathcal{L}(\mathbb{R}^p, W)$ is chosen such that $A_H = A + HC$ generates an exponentially stable semigroup on W, and $C_F = C + DF$, $B_H = B + HD$.

Of course, it is well known that stabilizability and detectability are related to the existence of solutions to algebraic Riccati equations and there is a complete theory for algebraic Riccati equations for the Pritchard–Salamon class.

THEOREM 5.3.2 ([130]). *If (C, A, B) belongs to the Pritchard–Salamon class, (A, B) is exponentially stabilizable on V and (C, A) is exponentially detectable on W, then for every $z \in Z = D_V(A)$, the following equation in Z^* has a unique nonnegative definite selfadjoint solution $Q \in \mathcal{L}(V, V^*)$*

$$(5.20)\qquad A^*Qz + QAz - QBB^*Qz + C^*Cz = 0.$$

Furthermore, $A - BB^*Q$ generates an exponentially stable C_0-semigroup on V.

We remark that we identify the duals of X, $\mathbb{R}^p$, and $\mathbb{R}^m$, but not those of V and W; there holds

$$(5.21)\qquad W \hookrightarrow X \hookrightarrow V, \qquad V^* \hookrightarrow X = X^* \hookrightarrow W^*.$$

Using the duality of the conditions (5.10) and (5.11), we obtain the following result on the filter algebraic Riccati equation.

COROLLARY 5.3.1. *Under the assumptions of Theorem 5.3.1 the following equation for all $y \in Z$ defined by* (5.9), *has a unique nonnegative definite, selfadjoint solution $P \in \mathcal{L}(W^*, W)$*

$$(5.22)\qquad APy + PA^*y - PC^*CPy + BB^*y = 0.$$

Furthermore, $A - PC^*C$ generates an exponentially stable C_0-semigroup on W.

As observed by [115] and [31] one can obtain normalized right and left coprime factorizations (see Definition 5.2.3) from (5.19) by choosing the feedbacks F and H from the control and filter Riccati equations; this extends to the Pritchard–Salamon class too.

COROLLARY 5.3.2 ([42]). *Under the assumptions of Lemma 5.3.6, if $D = 0$, then equations* (5.19) *yield a normalized doubly coprime factorization if F and H are chosen by*

$$(5.23)\qquad F = -B^*Q \in \mathcal{L}(V, \mathbb{R}^m); \qquad H = -PC^* \in \mathcal{L}(\mathbb{R}^p, W),$$

where Q and P are the solutions of the control and filter algebraic Riccati equations (5.20) *and* (5.22).

This type of factorization is exploited in §5.6 for the robust stabilization problem under normalized coprime factor perturbations.

For the special case when A generates an exponentially stable C_0-semigroup, equation (5.20) with $B = 0$ reduces to the Lyapunov equation for the *observability gramian* $Q_0 = \mathcal{C}^*\mathcal{C}$ and equation (5.22) with $C = 0$ reduces to the Lyapunov equation for the *controllability gramian* $P_0 = \mathcal{B}\mathcal{B}^*$. In [42] it is shown that the nonzero Hankel singular values of the Hankel operator Γ are equal to the square roots of the nonzero eigenvalues of P_0Q_o, just as in the finite-dimensional case [75].

Summarizing, we have identified the class of exponentially stabilizable and detectable Pritchard–Salamon systems to be one having properties very similar to those of finite-dimensional systems. In particular, they have all the properties required for a control synthesis along the lines advocated in [146] and [158], as we shall demonstrate in §§5.5–5.7.

5.4. General Distance Problems and L_∞-Approximation

The general distance or Nehari problem is an important step in solving H^∞-problems (see [75], [76], or [62]) and it has been studied by several authors in different forms ([1], [4], [85], [70], and [53]). The most general existence results is to be found in [4], but for our purposes the following special case suffices.

THEOREM 5.4.1 (NEHARI THEOREM). *Suppose G is a matrix-valued complex function in H^∞, which is the Laplace transform of a matrix-valued function $f \in L_1(0,\infty;\mathcal{L}(\mathbb{R}^m,\mathbb{R}^p))$. Then*

$$\inf_{J\in H_-^\infty} \|G - J\|_\infty = \|G\|_H, \tag{5.24}$$

where H_-^∞ is the space of matrix-valued functions of a complex variable which are analytic and bounded in $\operatorname{Re} s < 0$ *and $\|G\|_H$ is the induced L_2-norm of the Hankel operator corresponding to G. The infimum is attained by a $J \in H_-^\infty$.*

We recall that for our class of systems the Hankel operator Γ is defined by (5.17) and $\|G\|_H$ equals the largest singular value of Γ.

We shall see in §§5.6 and 5.7, that for H^∞-control problems for G belonging to the Pritchard–Salamon class, we require solutions J to (5.24) with the property that $J(-s) \in \mathcal{M}(\hat{\mathcal{A}}_-(0))$. This will not hold in general, but we do have some partial results in this direction. In particular, for the relaxed Nehari problem, which seeks all $J \in H_-^\infty$, which satisfy

$$\|G - J\|_\infty \leq \sigma \quad \text{for } \sigma > \|G\|_H \tag{5.25}$$

Curtain and Ran in [53] have obtained an explicit parametrization of all solutions in terms of realizations (C, A, B) of G.

THEOREM 5.4.2 ([53, THM. 3.4]). *If (C, A, B) belongs to the Pritchard–Salamon class and A generates an exponentially stable C_0-semigroup on V and*

W, then for $G(s) = C(s-A)^{-1}B$ all solutions to (5.25) are given by $J = G-F$, where

(5.26) $$F(s) = (\theta_{11}(s)K(s) + \theta_{12}(s))(\theta_{21}(s)K(s) + \theta_{22}(s))^{-1}$$

where $K \in H^{\infty}_{-}$, $\|K\|_{\infty} \leq 1$, and

(5.27) $$\begin{aligned}\theta_{11}(s) &= I + \sigma^{-2}C(s-A)^{-1}P_0\Phi C^*,\\ \theta_{12}(s) &= \sigma^{-1}C(s-A)^{-1}\Phi^* B,\\ \theta_{21}(s) &= -\sigma^{-2}B^*(s+A^*)^{-1}\Phi C^*,\\ \theta_{22}(s) &= \sigma^{-1}I - \sigma^{-3}B^*(s+A^*)^{-1}\Phi Q_0 B.\end{aligned}$$

P_0 and Q_0 are the controllability and observability gramians of (C, A, B), respectively, and $\Phi = (I - \sigma^{-2}Q_0P_0)^{-1}$.

From the formulas (5.26), (5.27) for the solution J and the fact that $Q \in \mathcal{L}(V, V^*)$ and $P_0 \in \mathcal{L}(W^*, W)$ from Theorem 5.3.1 and with $B = 0$ and Corollary 5.3.1 with $C = 0$, it is readily deduced that whenever $K(-s) \in \mathcal{M}(\hat{A}_-(0))$, so is $J(-s)$ [53, Lemma 3.6]. So the relaxed problem (5.25) has solutions with the required property, which is essential for control synthesis.

Another interesting class of infinite-dimensional systems was introduced in [85].

DEFINITION 5.4.1. (C, A, B) is said to be of *nuclear type* if its impulse response $h \in L_1 \cap L_2((0,\infty); \mathbb{R}^{p\times m})$ and its Hankel operator Γ is nuclear from $L_2(0,\infty;\mathbb{R}^m) \to L_2(0,\infty;\mathbb{R}^p)$

(5.28) $$(\Gamma u)(t) = \int_0^{\infty} h(t+s)u(s)ds.$$

Recall that Γ is nuclear if and only if the sum of its singular values is finite;

(5.29) $$\sum_{i=0}^{\infty} \sigma_i < \infty.$$

Nuclear systems always have a realization which is in the Pritchard–Salamon class, but the A operator may only generate a strongly stable semigroup ($\|e^{At}x\| \to 0$ as $t \to \infty$). Not all exponentially stable Pritchard–Salamon systems are nuclear; for example $G(s) = e^{-(s+1)/2}/(s+1+e^{-s-1})$ has an exponentially stable Pritchard–Salamon realization, but it is not nuclear [128].

Various tests for nuclearity exist, for example the following necessary and sufficient condition established in [34] for matrix-valued transfer functions $G \in H^{\infty}$

(5.30) $$\int_{-\infty}^{\infty}\int_0^{\infty} |G''(s)|\, dx\, dy < 0 \qquad (s = x+iy),$$

or alternatively that $G(s)$ possesses on $\mathbb{C}_0^+$ an expansion of the form

(5.31) $$G(s) = \sum_{n=1}^{\infty} \frac{c_n}{s-\lambda_n},$$

where

$$\sum_{n=1}^{\infty} \frac{|c_n|}{|\mathrm{Re}\,\lambda_n|} < \infty.$$

(5.31) is useful for p.d.e. models [41] and for delay systems more refined tests were developed in [128].

Examples of nuclear systems are Examples 5.2.1, 5.2.2, 5.2.4, and 5.3.1. Example 5.2.5 is not nuclear which can be seen by checking (5.30). It is interesting to see how nuclearity depends on how one models the sensors and actuators. If we suppose that either the actuator or the sensor has some dynamics, say the low pass filter $1/(s+\pi^2)$, then we obtain the modified transfer function for Example 5.2.5

$$G(s) = \frac{1}{s+\pi^2} \frac{\sinh \sqrt{s} x_0}{\sinh \sqrt{s}}$$

$$= \frac{1}{s+\pi^2} \left(\sum_{n=1}^{\infty} \frac{(-1)^n n \sin n\pi x_0}{\pi(n^2-1)} \right) - \sum_{n=1}^{\infty} \frac{(-1)^n n \sin n\pi x_0}{\pi(n^2-1)(s+(n\pi)^2)}$$

and that this is nuclear can be easily verified using the condition (5.31). This type of analysis applies to other transfer functions that have only finitely many unstable poles, but whose transfer function is not nuclear. In other words, by introducing some dynamics into the loop, either by modeling or by design (precompensation), we obtain a "nicer" transfer function, which is easier to control and approximate. This is, of course, standard engineering practice [83].

It is known that the Nehari problem for the case that G is nuclear does have smooth solutions J, such that $J(-s) \in \mathcal{M}(\hat{\mathcal{A}}(0))$ [42]. However, it is not known whether $J(-s) \in \mathcal{M}(\hat{\mathcal{A}}_-(0))$ in general.

The real interest in nuclear systems lies in their approximation properties, which have been extensively studied in [166], [88], [128], [85], and [41]. As in finite dimensions, good L_∞-approximations are obtained by truncating balanced realizations, which exist for a very general class of infinite-dimensional systems [48], [121]. *Balanced realizations* (C_b, A_b, B_b) for a transfer function G have the property that $G(s) = C_b(s-A_b)^{-1}B_b$ and the controllability and observability gramians are both equal to $\Sigma = \mathrm{diag}(\sigma_1, \sigma_2, \cdots)$, where σ_i are the Hankel singular values of G. If the singular values of G are simple and G is of nuclear type, then in [85] it was proved that the nth order truncation of the balanced realization (C_b^n, A_b^n, B_b^n) approximates G with the following error estimate

$$\|G(s) - C_b^n(s-A_b^n)^{-1}B_b^n\|_\infty \leq 2 \sum_{i=n+1}^{\infty} \sigma_i. \tag{5.32}$$

Even better L_∞-approximations are obtained by the so-called *optimal Hankel norm approximations.* These approximants $\hat{G}_n$ are solutions of problem

$$\inf_{G_n} \|G - G_n\|_H \tag{5.33}$$

where $\hat{G}_n$ has McMillan degree $\leq n$. These problems have an extensive literature [1], [4], but the main interest for control applications is that for the case when G is nuclear, $\|G - \hat{G}_n\|_\infty \sim \sum_{i=n+1}^{\infty} \sigma_i$ or about half that for the truncated balanced approximations. This and several other types of error estimates can be found in [81], [85] and numerical techniques for calculating these and other good L_∞-approximations for nuclear systems can be found in [88], [128], and [41]. The main advantage of truncated balanced and optimal Hankel norm approximations over other L_∞-approximation schemes lies in the fact that for systems of nuclear type one can calculate a priori error bounds, which is extremely important for robust controller design (see [82], [47], [49], [50], [22], [23], [25], [41], and §5.5).

Even for nonnuclear systems truncated balanced and optimal Hankel norm approximations often yield good approximations. This phenomenon is now well understood for delay systems and [89] gives a comprehensive treatment of asymptotic estimates for the singular values and L_∞-approximations of delay systems with a compact Hankel operator: useful comparisons are made between the L_∞-errors of balanced, optimal Hankel norm and Padé approximations for general delay systems (L_2-approximations of delay systems are also considered). For other systems (such as those in Examples 5.2.4–5.2.7) only the nuclear case is well understood, although the results in [120] might prove useful here.

Recently a Fourier transform approach to L_∞-approximation was suggested in [90] for a class of infinite-dimensional systems which is close to the nuclear class. The approximate model can be directly computed using a fast Fourier transform type algorithm which only requires experimental data of the original transfer function. This can be an advantage over the balanced and optimal Hankel norm approximations which require an analytic expression for the transfer function, but then the latter in turn have the advantage that they give a priori bounds on the L_∞-error, which is so important for robust controller design.

5.5. Robust Stabilization under Additive Perturbations

There exist many theories for designing stabilizing controllers, both finite and infinite-dimensional, either using state-space approaches [2], [3], [17], [19], [32], [35], [59], [97] or frequency domain approaches [13], [14], [30], [99], [106], [116] to mention only a fraction of the known results.

Since in practice one nearly always has only an approximate model of a system, it is important to know that the controller one designs for this *nominal* model will also stabilize other systems which are "close" to the nominal model; this is the *robustness issue.* There are many possible ways of defining robustness and here we consider one in which the perturbations are assumed to occur additively. The aim is to design controllers which have an a priori guaranteed robustness margin. In [82] Glover developed a theory for such controllers for finite-dimensional systems and this was extended in [47]–

[50] to the Callier–Desoer class introduced in §5.2. Here we present a somewhat simplified account of the theory from [50], which considers robustness of the controller system of Fig. 5.1 under unstructured, additive, infinite-dimensional perturbations, Δ.

The precise conditions on G and Δ are assumed to be the following:

(5.34) The nominal system $G \in \mathcal{M}(\hat{\mathcal{B}}(0))$ is strictly proper and has no poles on the imaginary axis;

(5.35) The permissible perturbations $\Delta \in \mathcal{M}(\hat{\mathcal{B}}(0))$ are strictly proper and are such that Δ and $\Delta + G$ have equal numbers of poles in $\mathrm{Re}\, s \geq 0$;

and we are interested in how large the perturbations Δ can be without destroying the stability of the closed-loop system of Fig. 5.2.

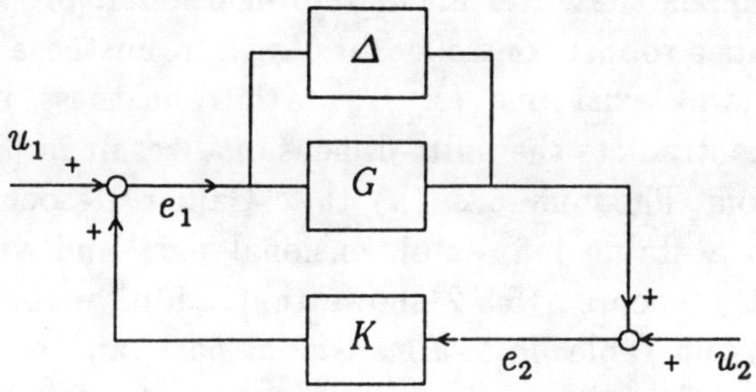

FIG. 5.2. *Additive robust stability.*

Notice that our admissible perturbations allow for arbitrarily large delays, $\tau > 0$, such as $\Delta(s) = Me^{-\tau s}$; however, we shall presently impose a restriction on the magnitude, M, of its L_∞-norm.

DEFINITION 5.5.1. The feedback system (G, K) of Fig. 5.2 with $G, K \in \mathcal{M}(\hat{\mathcal{B}}(0))$ is said to be *additively robustly stable with robustness margin* ϵ if the feedback system of Fig. 5.2 is input-output stable for all G and Δ satisfying (5.34) and (5.35) and $\|\Delta\|_\infty < \epsilon$.

Robust stabilizability was studied in [33] where using Nyquist arguments they proved the following theorem on robust stabilization under additive perturbations.

THEOREM 5.5.1. *If* $G, K \in \mathcal{M}(\hat{\mathcal{B}}(0))$ *and* (5.34), (5.35) *are satisfied,* (G, K) *is additively robustly stable with robustness margin* $\epsilon > 0$ *if and only if* (G, K) *is input-output stable and*

$$\|K(I - GK)^{-1}\|_\infty \leq \frac{1}{\epsilon}. \tag{5.36}$$

This theorem can be used to compare the robustness of different controllers as in [23], but more interesting is that it can be used to obtain the following powerful result on the maximal robustness margin.

THEOREM 5.5.2 ([50]). *If* $G, K \in \mathcal{M}(\hat{\mathcal{B}}(0))$ *and* (5.34) *is satisfied, given an* $\epsilon > 0$ *there exists a compensator* K *which stabilizes* $G + \Delta$ *for all* Δ *satisfying* (5.35) *and* $\|\Delta\|_\infty < \epsilon$ *if and only if*

$$\sigma_{\min}(G_u) \geq \epsilon, \tag{5.37}$$

where $\sigma_{\min}(G_u)$ *is the smallest Hankel singular value of the finite-dimensional unstable part of* G.

In fact, only the sufficient part is proved in [50]; however, the necessity part follows readily from that for the finite-dimensional case [82]. This is because the infinite-dimensional controllers K_u for G_u and K for G are in one-to-one correspondence: $K = K_u(I + G_s K_u)^{-1}$ and $K_u = (I + KG_s)^{-1}K$. Moreover, they have the same robustness margin given by (5.36):

$$K(I - GK)^{-1} = K_u(I - G_u K_u)^{-1}. \tag{5.38}$$

If the G in (5.36) is finite-dimensional (say G_u), then the infimum over all K in $\mathcal{M}(\hat{\mathcal{B}}(0))$ equals that over all finite-dimensional, proper controllers K. Thus if there exists a robust controller for G with robustness margin exceeding $\sigma_{\min}(G_u)$, there will exist one for G_u with robustness margin exceeding $\sigma_{\min}(G_u)$. This contradicts the finite-dimensional result in [82].

We recall from Theorem 5.2.1(b) that Callier–Desoer systems can be decomposed into a stable infinite-dimensional part and an unstable finite-dimensional part. Theorem 5.5.2 shows that additive robustness is purely a property of this unstable *finite-dimensional* part, and this is usually more accurately modeled than the infinite-dimensional part. Moreover, $\sigma_{\min}(G_u)$ can be easily calculated a priori yielding valuable information about the maximum achievable robustness for a nominal model G. The same theorem can be applied to $G(s - \beta)$ giving information about robustness with respect to an arbitrary stability margin β.

In addition to the existence result in [50], there are also explicit formulas for a controller K which achieves the maximum robustness margin $\sigma_{\min}(G_u)$. These depend on the stable part of G, G_s, and so this K is *infinite-dimensional*, which is often undesirable for applications. However, if one replaces G_s by a finite-dimensional L_∞-approximation, G_s^k, say, then one can design a strictly proper finite-dimensional controller (of the order of k plus the McMillan degree of G_u) which has a robustness margin of at least $\sigma_{\min}(G_u) - \|G_s - G_s^k\|_\infty$. This controller is robust under perturbations Δ which are proper, but not necessarily strictly proper (as required in (5.35)). So in combination with a theory for good L_∞-approximations (for example, that described in §5.4), we have a practical design technique for additively robust finite-dimensional controllers for infinite-dimensional systems of the Callier–Desoer class [41]. Finite-dimensional controllers for infinite-dimensional systems based on reduced-order models have appeared elsewhere in the literature [3], but [47], [50] were the first results to give a priori estimates for the achievable robustness and to clarify the role of approximation in controller design.

In particular, Theorem 5.5.2 explains the mysteries surrounding "*spillover*" which sometimes occurs in controller design based on reduced-order models [3]. In reduced-order model design, it is usual to design a finite-dimensional controller K_f based on some low-order approximation G_f of the infinite-dimensional system G and then hope that K_f will also stabilize G. Theorems 5.5.1 and 5.5.2 now applied to G_f tell us that K_f will stabilize G if it stabilizes G_f, G and G_f have the same number of unstable poles, and

$$\|G - G_f\|_\infty \leq \|K_f(I - G_f K_f)^{-1}\|_\infty^{-1}. \tag{5.39}$$

An intelligent choice for K_f would be the maximally robust controller for G_f, for then Theorem 5.5.2 tells us that K_f stabilizes G if

$$\|G - G_f\|_\infty \leq \sigma_{\min}(G_f^u), \tag{5.40}$$

where G_f^u is the unstable part of G_f. Inequality (5.39) gives an a priori criterion for choosing the reduced order model G_f for G. So we see that spillover only arises if the error of the reduced-order model exceeds $\sigma_{\min}(G_f^u)$ and this can be calculated a priori from the chosen reduced-order model. Furthermore, Theorem 5.5.2 clarifies the trade-off between the robustness margin, the order of the controller, and the order of the reduced-order model (see [22]).

We remark that the only requirement of the theory is that the system be strictly proper and belong to the Callier–Desoer class and so it applies to Examples 5.2.1, 5.2.2, 5.2.4, 5.2.5 (heat equation with Dirichlet boundary control) and in fact to all exponentially stabilizable and detectable Pritchard–Salamon systems (Lemma 5.3.3). This Pritchard–Salamon class has the attractive feature that input-output stability implies exponential stability of a detectable and stabilizable realization (cf. Theorem 5.3.1). Moreover, we recall that strictly proper transfer matrices in the $\hat{\mathcal{A}}_-(0)$ class can be approximated in the L_∞-norm by a stable rational transfer function [115a] and so we can obtain finite-dimensional robust stabilizing controllers for all these examples. The theory has been applied to delay systems in [47] and in comparison studies on various beam models in [23]–[26]. We illustrate the approach by means of the beam Example 5.3.1.

Example 5.5.1. In Example 5.3.1 we established that the beam model belonged to the Pritchard–Salamon class, and we obtained a series expansion for its transfer function, which is easily shown to be equal to

$$\begin{aligned} G(s) &= \frac{1}{\rho a}\begin{bmatrix} \dfrac{1}{2s^2} & 0 \\ 0 & \dfrac{3}{2s^2} \end{bmatrix} \\ &\quad + \frac{1}{\rho a}\sum_{n=3}^{\infty}\begin{bmatrix} v_n^2(0) & 0 \\ 0 & v_{nx}^2(0) \end{bmatrix} / (s^2 + \alpha_1\lambda_n^4 s + \alpha_2\lambda_n^4), \end{aligned} \tag{5.41}$$

where λ_n and $v_n(x)$ are the eigenvalues and eigenfunctions of the stiffness operator A, as explained in Example 5.3.1. Clearly, $G(s)$ is strictly proper and it is in $\mathcal{M}(\hat{\mathcal{B}}(0))$; however, it has two poles on the imaginary axis, and

(5.34) is not satisfied. This slight technical difficulty disappears if we demand a stability margin of $\beta = 0.0065$, say. We then apply our control design to the transfer function $G_\beta(s) = G(s-\beta)$ which satisfies assumptions (5.34) and (5.35). G_β then has six "unstable" poles and the maximum robustness margin is equal to the smallest singular value of the finite-dimensional system

$$G_u^\beta(s) = \frac{1}{\rho a}\begin{bmatrix} \dfrac{1}{2(s-\beta)^2} & 0 \\ 0 & \dfrac{3}{2(s-\beta)^2} \end{bmatrix} \tag{5.42}$$

$$+\frac{1}{\rho a}\sum_{n=3}^{4}\begin{bmatrix} v_n^2(0) & 0 \\ 0 & v_{nx}^2(0) \end{bmatrix} / (s^2 + \alpha_1\lambda_n^4 s + \alpha_2\lambda_n^4).$$

This is readily calculated using the matrix manipulation package PCMATLAB to be 1.5855. As expected, the robustness margin decreases with increasing stability margin, e.g., for $\beta = .007$ and 0.008 the maximum robustness margins are .4784 and .0963, respectively. To design a finite-dimensional controller, we need to approximate the "unstable" part (with respect to β), i.e.,

$$G_s^\beta(s) = \frac{1}{\rho a}\sum_{n=5}^{\infty}\begin{bmatrix} v_n^2(0) & 0 \\ 0 & v_{nx}^2(0) \end{bmatrix} / ((s-\beta)^2 + \alpha_1\lambda_n^4(s-\beta) + \alpha_2\lambda_n^4). \tag{5.43}$$

In fact $\|G_s^\beta\|_\infty$ is extremely small in this example and approximating the "unstable" part by zero, we obtain a guaranteed robustness margin of 1.5865. This can be achieved by a fifth-order controller which is designed on the finite-dimensional approximation, again using PCMATLAB. Of course, it is essential to have some idea what this robustness margin means in terms of physical characteristics of the beam, which may be difficult to model exactly. In particular, it is important to know if a robust controller that has a robustness margin of 1.5865 will also stabilize the same model for different values of the parameters α_1 and α_2 or the same beam modeled by a different p.d.e. or with slightly different inputs or outputs. These and similar questions concerning the applicability of the theory of additively robust stabilizers to structure perturbations were addressed in [20].

The fact that all the above calculations are standard *finite-dimensional* ones that can be performed using PCMATLAB [129] is not a special feature of this example. It is a valid comment for any delay or p.d.e. system that has a transfer function in partial fraction form. Methods for finding such an expression for delay systems are given in [166].

This section comprised a theory for designing additively robust controllers for strictly proper systems belonging to the Callier–Desoer class having no poles on the imaginary axis. In fact, we have presented a simplified version of the original results in [47], [50], where rational weighting functions were also included. These are often introduced to induce certain desired frequency behavior in the closed-loop system and they give extra flexibility in the design of the controllers. As we have seen in Example 5.5.1 we can handle poles on the imaginary axis by introducing a stability margin β; such stability enhancement is common practice. Recently, in [109] Logemann extended these results to a larger class of transfer functions, those that can be represented as the sum of an unstable finite-dimensional part and a stable infinite-dimensional part which is in $H^\infty(-\epsilon)$, the class of functions of a complex-variable which are analytic and bounded in $\operatorname{Re} s > -\epsilon$ for some $\epsilon > 0$. If one wishes to design finite-dimensional controllers for this class then the approximation step may prove difficult and limits the robustness margin. Alternatively, it might prove useful to modify the original transfer function by means of a smoothing precompensator so as to obtain one in the Callier–Desoer class. This device can also be used to obtain a strictly proper transfer function [83].

So the essential limitations on the applicability of this robustness theory which remain are that the system can have only finitely many unstable poles and the class of admissible perturbations should satisfy (5.35), which means that they are additive, unstructured, and the perturbed system should have the same number of unstable poles as the unperturbed system. In §5.6 we present another theory for unstructured perturbations which removes this last restriction on the number of unstable poles in the perturbed system.

5.6. Robust Stabilization under Normalized Coprime Factor Perturbations

In §5.5 we outlined a theory for robust stability with respect to unstructured, stable, additive perturbations. Here we consider robust stabilization with respect to additive, stable perturbations to a coprime factorization of the system, which allows for unstable perturbations via the denominator factor. This type of robustness was advocated by Vidyasagar in [146], [147], where he showed that this family of perturbations is particularly appropriate for feedback systems analysis. Recently in [87], [114] a very pleasing, explicit solution to this problem was obtained for normalized coprime factorizations of finite-dimensional systems. At the same time, it provides one of the few examples of an H^∞-problem that can be solved completely and explicitly in terms of the original system parameters. By synthesizing the frequency domain description of §5.2 and the state-space description of §5.3 it is not too difficult to see how one can generalize the results of [87], [114] to a class of infinite-dimensional systems, namely, the Pritchard–Salamon class (C, A, B) with the property that (A, B) is exponentially stabilizable on V and (C, A) is exponentially detectable on W; this was done in [42]. The difficult steps arise in bridging the gap between the *algebraic* control synthesis and the *analytic* general distance problem and in obtaining useful numerical algorithms. The following is a brief account of the article [42], which achieves the first step of obtaining a theoretical solution to this problem.

We are interested in the robustness of input-output stability for exponentially detectable and stabilizable Pritchard–Salamon systems under certain perturbations with respect to a given normalized left coprime factorization $G = \tilde{M}^{-1}\tilde{N}$ (cf. Definition 5.2.4). The perturbations we consider are of the form

$$G_\Delta = (\tilde{M} + \Delta_M)^{-1}(\tilde{N} + \Delta_N), \tag{5.44}$$

where Δ_N, Δ_M are unknown (stable) transfer functions in $\mathcal{M}(\hat{\mathcal{A}}_-(0))$ which represent the uncertainty in our nominal system model. Notice that there are no restrictions on the number of unstable poles in G_Δ, in contrast to the additive perturbation theory given in §5.5 (cf. (5.35)).

The robust design objective is to find a feedback controller $K \in \mathcal{M}(\hat{\mathcal{B}}(0))$ which stabilizes not only the nominal system G, but also the family of perturbed systems defined by

(5.45)

$$\mathcal{G}_\epsilon = \{G_\Delta = (\tilde{M} + \Delta_M)^{-1}(\tilde{N} + \Delta_N) \text{ such that } \Delta_M, \Delta_N \in \mathcal{M}(\hat{\mathcal{A}}_-(0)) \text{ and } \|[\Delta_M, \Delta_N]\|_\infty < \epsilon\} \quad \text{(see Fig. 5.3)}$$

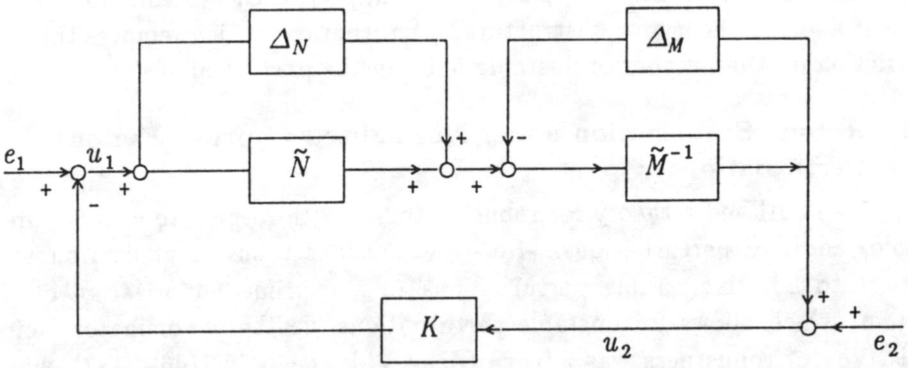

FIG. 5.3. *Left coprime factorization perturbations.*

This leads to the following definition of factor robust stability.

DEFINITION 5.6.1. Suppose that the system $G \in \mathcal{M}(\hat{\mathcal{B}}(0))$ has the left normalized coprime factorization of Definition 5.2.4. Then the feedback system $(\tilde{M}, \tilde{N}, K, \epsilon)$ of Fig. 5.3 is *factor robustly stable* if and only if (G_Δ, K) is input-output stable for all $G_\Delta \in \mathcal{G}_\epsilon$. If there exists a K such that $(\tilde{M}, \tilde{N}, K, \epsilon)$ is factor robustly stable, then (G, ϵ) is said to be factor robustly stabilizable with *factor robustness margin* ϵ.

(Input-output stability is defined in Definition 5.2.5.)

Appealing again to the results of [33] applied to Fig. 5.3 (cf. Theorem 5.5.1), we obtain the following characterization of factor robust stabilizability.

LEMMA 5.6.1. *Suppose that $G \in \mathcal{M}(\hat{\mathcal{B}}(0))$ has the normalized left coprime factorization $G = \tilde{M}^{-1}\tilde{N}$, then (G, ϵ) is factor robustly stabilizable if and only if*

$$\inf_K \left\| \begin{bmatrix} K \\ I \end{bmatrix} (I - GK)^{-1}[G \; I]_\infty \right\| \leq \frac{1}{\epsilon}, \tag{5.46}$$

where the infimum is taken over all stabilizing controllers $K \in \mathcal{M}(\hat{\mathcal{B}}(0))$.

Inequality (5.46) has the form of an H^∞-optimization problem and it can be converted to the general formulation of

$$\inf_K \|\mathcal{F}_L(P, K)\|_\infty \leq \frac{1}{\epsilon}, \tag{5.47}$$

where

$$P = \begin{bmatrix} P_{11} & P_{12} \\ P_{21} & P_{22} \end{bmatrix} = \begin{bmatrix} 0 & I \\ \tilde{M}^{-1} & G \\ \tilde{M}^{-1} & G \end{bmatrix} \tag{5.48}$$

and $\mathcal{F}_L(P, K)$ denotes a *linear fractional transformation* defined by

$$\mathcal{F}_L(P, K) = P_{11} + P_{12}K(I - P_{22}K)^{-1}P_{21}. \tag{5.49}$$

Since G possesses a *doubly normalized coprime factorization* (Corollary 5.3.2), we can reduce (5.49) to a simpler form using the Youla parametrization of stabilizing controllers from Theorem 5.2.2:

$$\mathcal{F}_L(P, K) = \begin{bmatrix} Y \\ X \end{bmatrix} + \begin{bmatrix} M \\ N \end{bmatrix} J, \tag{5.50}$$

where $J \in \mathcal{M}(\hat{\mathcal{A}}_-(0))$ is the free parameter from (5.6) and X, Y, M, N are given by (5.19) with F and H as in (5.23). The parametrization (5.50) is in fact valid for all left coprime factorizations of G and a similar parametrization in terms of right-coprime factorizations can be found in [147].

So we have reduced our robust stabilization problem to the following model matching problem

$$\inf_{J \in \mathcal{M}(\hat{\mathcal{A}}_-(0))} \left\| \begin{bmatrix} Y \\ X \end{bmatrix} + \begin{bmatrix} M \\ N \end{bmatrix} J \right\|_\infty, \tag{5.51}$$

which in turn can be reduced to the general distance problem

$$\inf_{J \in \mathcal{M}(\hat{\mathcal{A}}_-(0))} \left\| \begin{bmatrix} R_1 + J \\ I \end{bmatrix} \right\|_\infty, \tag{5.52}$$

where R_1 is the antistable system given by

$$R_1^t(-s) = C(I + PQ)(s - A + BB^*Q)^{-1}B. \tag{5.53}$$

Here we are confronted with a problem which does not arise in finite-dimensions. Because we define stable systems to be in $\mathcal{M}(\hat{\mathcal{A}}_-(0))$, we seek solutions to the general distance problem (5.52) in the class $\mathcal{M}(\hat{\mathcal{A}}_-(0))$, whereas the general theory gives solutions in the much larger class H^∞.

This problem has already been discussed in §5.4 and Theorem 5.4.1 gives the solution to the related general problem over H^∞:

$$(5.54)\qquad \inf_{J\in H^\infty}\left\|\begin{bmatrix} R_1+J \\ I\end{bmatrix}\right\|_\infty = (1+\|R_1(-s)\|_H^2)^{\frac{1}{2}} = (1+\lambda_{\max}(PQ))^{\frac{1}{2}}$$

where Q and P are the solutions to the control and filter Riccati equations (5.20), (5.22).

In fact, general theory shows that $\|R_1(-s)\|_H^2 = \lambda_{\max}(P_0Q_0)$, where P_0 and Q_0 are the controllability and observability gramians of R_1 (see the remark after Corollary 5.3.2) and it just so happens that $P_0Q_0 = PQ$.

It is not known when the solutions to (5.52) and (5.54) agree, but in any case (5.54) gives an upper bound to the achievable factor robustness margin of

$$(5.55)\qquad \epsilon_{\max} \le [1+\lambda_{\max}(PQ)]^{\frac{1}{2}} = (1-\|[\tilde{N},\tilde{M}]\|_H^2)^{\frac{1}{2}}.$$

We recall from §5.4, Theorem 5.4.2, that there always exists a $J\in\mathcal{M}(\hat{A}_-(0))$ which will almost achieve this bound $\epsilon_{\max}$. This approximating solution J can be used to synthesize a controller using the Youla formulas (5.5) with X, Y, M, etc., given by (5.19) and (5.23). In this way we can obtain explicit formulas for nearly optimal factor robust controllers just as in [114] and thus we obtain a complete generalization of the results of [114] to the Pritchard–Salamon class of systems which are exponentially stabilizable and detectable. Of course, by stability here we mean the input-output stability of Definition 5.2.5. However, it is important to note that since the Pritchard–Salamon class (C,A,B) is exponentially stabilizable and detectable, the input-output stable closed-loop system will be exponentially stable whenever the controller also has an exponentially stabilizable and detectable Pritchard–Salamon realization [39].

Although the above is a nice *theoretical existence* result, it is very far from becoming a useful controller design. First we should notice that X, Y, M, etc., in (5.19) and (5.23) are *infinite-dimensional* and so the controller will be infinite-dimensional too, which is undesirable. Moreover, everything depends on P and Q, the solutions of infinite-dimensional Riccati equations. Fortunately, it is always possible to approximate the system a priori and reduce all the calculations to finite-dimensional ones as for the additive robust controllers of §5.5. To justify this we make use of the new result in [78], where they solve the abstract robust stabilizability problem for a larger class of transfer matrices, $G=\tilde{M}^{-1}\tilde{N}$, where $\tilde{M}$ and $\tilde{N}$ are in $\mathcal{M}(H^\infty)$. It is a purely frequency domain approach and uses the notion of the gap-metric. For simplicity we shall state their results restricted to our class of systems.

DEFINITION 5.6.2. Let G_1 and G_2 have the normalized left coprime factorizations $G_i=\tilde{M}_i^{-1}\tilde{N}_i$; $i=1,2$. The *directed T-gap* is defined by

$$(5.56)\qquad \vec{\delta}_t(G_1,G_2) = \inf_{Q\in H^\infty}\|[\tilde{M}_1,\tilde{N}_1]-Q[\tilde{M}_2,\tilde{N}_2]\|_\infty$$

and the T-gap by

$$\delta_T(G_1, G_2 = \max\{\vec{\delta}_T(G_1, G_2), \vec{\delta}_T(G_2, G_1)\}. \tag{5.57}$$

In general, $\vec{\delta}_T(G_1, G_2) \neq \vec{\delta}_T(G_1, G_2)$, but if $\delta_T(G_1, G_2) < 1$, then they are both equal. Reference [78] contains many interesting results about robust controllers for finite- and infinite-dimensional systems, but the following one is of most interest to our situation.

THEOREM 5.6.1. *Consider a system G with the normalized coprime factorization, $G = \tilde{M}^{-1}\tilde{N}$ and let K be any controller which stabilizes G. For real ϵ with $0 < \epsilon \leq 1$, the following statements are equivalent:*

(a) *K stabilizes all $G_1 = (\tilde{M} + \Delta_M)^{-1}(\tilde{N} + \Delta_N)$, where Δ_M, $\Delta_N \in \mathcal{M}(H^\infty)$ and $\|[\Delta_M, \Delta_N]\|_\infty < \epsilon$.*

(b) *K stabilizes all G_1 with $\vec{\delta}_T(G_1, G_2) < \epsilon$.*

(c) *K stabilizes all G_1 with $\delta_T(G_1, G_2) < \epsilon$.*

Furthermore, the maximum factor robustness margin is given by

$$\epsilon_{\max} = (1 - \|[\tilde{M}, \tilde{N}]\|_H^2)^{\frac{1}{2}}. \tag{5.58}$$

We remark that by allowing the wider class of perturbations we now have equality in (5.58). We now use Theorem 5.6.1 to prove the following useful approximation lemma.

LEMMA 5.6.2. *Suppose that $G = G_f + G_s$, where G_f is rational and contains all the unstable modes and $G_s \in \mathcal{M}(\hat{\mathcal{A}}_-(0))$ and*

$$\|G_s\|_\infty < \mu << \epsilon_{\max} < 1. \tag{5.59}$$

Then if K_f stabilizes G_f with factor robustness margin ϵ, it stabilizes G with factor robustness margin at least equal to $\epsilon - \mu$.

Proof. If G_f has the normalized left-coprime factorization $G_f = \tilde{M}_f^{-1}\tilde{N}_f$, then G has the following left-coprime factorization

$$G = \tilde{M}_f^{-1}(\tilde{N}_f + \tilde{M}_f G_s).$$

It will not be normalized, but $G = \tilde{M}^{-1}\tilde{N}$ will be a normalized left-coprime factorization, where $U\tilde{M} = \tilde{M}_f$ and $U\tilde{N} = \tilde{N}_f + \tilde{M}_f G_f$ for some U invertible over $\mathcal{M}(\hat{\mathcal{A}}_-(0))$. We now estimate the directed T-gap between G_f and G

$$\begin{aligned}\vec{\delta}_T(G_f, G) &= \inf_{Q \in H^\infty} \|[\tilde{M}_f, \tilde{N}_f] - Q[\tilde{M}, \tilde{N}]\|_\infty \\ &\leq \|[\tilde{M}_f, \tilde{N}_f] - U[\tilde{M}, \tilde{N}]\|_\infty \\ &= \|\tilde{M}_f G_s\|_\infty \\ &\leq \|G_s\|_\infty \quad \text{by (5.4)}.\end{aligned}$$

Interchanging the roles of G and G_f, if $G = \tilde{M}^{-1}\tilde{N}$ is a normalized left-coprime factorization for G, $G_f = \tilde{M}^{-1}(\tilde{N} - \tilde{M}G_s)$ is a left-coprime one for G_f.

As above, we obtain the inequality

$$\vec{\delta}_T(G_f, G) \le \|G_s\|_\infty,$$

$\|G_s\|_\infty < 1$ and so the directed gaps are equal and

$$\delta_T(G, G_f) \le \|G_s\|_\infty. \tag{5.60}$$

Suppose now that K_f stabilizes G_f with factor robustness margin ϵ. From Theorem 5.6.1, it stabilizes all G_Δ such that $\delta_T(G_f, G_\Delta) < \epsilon$. Now

$$\begin{aligned}\delta_T(G_f, G_\Delta) &\le \delta_T(G_f, G) + \delta_T(G, G_\Delta)\\ &\le \|G_s\|_\infty + \delta_T(G, G_\Delta) \quad \text{by (5.60)}\\ &< \mu + \delta_T(G, G_\Delta) \quad \text{by (5.59)}\\ &< \epsilon\end{aligned}$$

provided that $\delta_T(G, G_\Delta) < \epsilon - \mu$. So K_f stabilizes all G_Δ with the property $\delta_T(G, G_\Delta) < \epsilon - \mu$, and appealing to Theorem 5.6.1 completes the proof.

We illustrate the application of this theory to our beam Example 5.3.1.

Example 5.6.1. Again we take the transfer matrix (5.41) as our starting point and using the same nominal values as in Example 5.5.1, we see that taking the tenth-order modal approximation as our finite-dimensional approximation, G_f, we incur an error $\|G_s\|_\infty \sim 0.02$. The maximum factor robustness margin for G_f is 0.3827 and so if we design a maximum robust controller K_f for G_f it also serves as a robust controller for the original infinite-dimensional system with factor robustness margin of at least 0.3627. Again all the calculations are finite-dimensional ones and can be performed using PCMATLAB. In [26] the dependence of the factor robustness margin on the parameters α_1 and α_2 is investigated and it is found that the robustness to variations in these parameters is significantly better than for the additively robust case (from §5.5). However, if we require not only stability, but stability enhancement, then we find that the factor robustness margin decreases as the stability margin increases. For a stability margin of $\beta = 0.0065$ this type of robust controller is only slightly better than the additively robust one with respect to variations is the parameters α_1 and α_2. For more details on these aspects, see [26] and [24].

References [126] and [127] present a different approach to robust controller design with respect to coprime factor perturbations which is applicable to delay systems.

Although we have given only an explicit solution (in terms of Riccati equations) to the robust stabilization problem with respect to stable factor perturbations for the restricted class of exponentially stabilizable and detectable Pritchard–Salamon systems, we can solve the practical problem for a much larger class. By practical we mean that we can actually design robust controllers with an a priori factor robustness margin. Theorem 5.6.1 shows that all we need to do is express our system $G = (\tilde{M}_f + \Delta_M)^{-1}(\tilde{N}_f + \Delta_N)$, where

$\tilde{M}_f^{-1}\tilde{N}_f = G_f$ is a normalized coprime factorization of the finite-dimensional system G_f and $\|[\Delta_M, \Delta_N]\|_\infty$ is much less than the maximum factor robustness margin of G_f. In particular, Lemma 5.6.2 gives a procedure for doing this for any system which can be decomposed as the sum of a finite-dimensional part and an infinite-dimensional stable part of sufficiently small L_∞-norm. An example of this type is the heat equation with Dirichlet boundary control. Summarizing, we see that we can design finite-dimensional factor robust controllers with an a priori robustness margin for a larger class of systems than those considered in §5.5; we can allow for poles on the imaginary axis and our class of perturbations permits changes in the number of unstable poles. Just as in §5.5, the control design amounts to an approximation step followed by a finite-dimensional control design.

In [78] the same factor robustness problem is solved for the systems belonging to the quotient field of $\mathcal{M}(H^\infty)$; i.e., G has a factorization $G = \tilde{M}^{-1}\tilde{N}$ where $\tilde{M}, \tilde{N} \in \mathcal{M}(H^\infty)$. This is probably the largest class of systems one would ever want to consider and it allows for infinitely many unstable poles and improper systems. One can define coprime factorizations analogously to Definition 5.2.4, but not all systems in the quotient field will have a coprime factorization. A very important necessary and sufficient condition for G to have a coprime factorization is that it be stabilizable (in the sense of Definition 5.2.5, [96], [139]). Moreover, if G has a coprime factorization, it always has a normalized one [163], and so the factor robustness problem is well posed for the quotient field of $\mathcal{M}(H^\infty)$. A key result from [78] states that Theorem 5.6.1(a) and (b) hold for this class and if G is input-output stabilizable it has a maximum robustness margin given by (5.58). Of course, these results are at a very abstract level and the next step is to develop techniques for finding normalized coprime factorizations. This has been done in [80] for a class of single-input, single-output systems. The contribution described in this section comes down to showing how to do this for a certain class, and for the exponentially stabilizable and detectable Pritchard–Salamon class, explicit state-space formulas are given for the solutions. Recent work in [54] illustrates that a similar approach is also feasible for plants with infinitely many unstable poles; these systems are not exponentially stabilizable or detectable, but a robustness property holds. This gives hope for giving a state-space interpretation of robustness for a wider class of systems and it seems a promising area of research.

Having stressed the attractiveness of the synthesis of state-space and frequency domain techniques, we should also point out that for some systems a purely frequency domain approach does offer advantages. Some preliminary work on coprime factorizations using a frequency domain approach can be found in [15], but more research needs to be done on this problem. Other frequency domain results related to robust stabilizability with respect to factor perturbations using the gap metric and the graph metric introduced in [146] have been obtained for the Callier–Desoer class of systems in [147], [161]–[165].

These have the disadvantage that the results so far are only theoretical and the computational aspects still need further investigation. For this direction we have already mentioned [80].

5.7. H∞-Control Theory for Infinite-Dimensional Systems: Open Problems and Promising Directions

H^∞-control problems are about the design of controllers to meet performance specifications such as robustness of stability to perturbations, minimizing sensitivity of outputs to certain input channels, tracking reference signals, bandwidth specifications and others that are naturally formulated in the frequency domain. Over the past few years, great success has been made in developing an elegant mathematical theory for such problems for finite-dimensional systems [75], [62], [76] and current research in this area dominates the control scene [67], [68], [73], [74], [86], [100], [125], [159], [160], [84]. Since the same performance specifications are desirable for infinite-dimensional systems, it is natural to study H^∞-control problems in an infinite-dimensional setting.

First we examine the H^∞-problem for the Pritchard–Salamon class of systems which are exponentially stabilizable and detectable. For this class we solved explicitly a particular H^∞-problem in §5.6. If one is familiar with the account of H^∞-control theory from [75], [76], or [62], it is clear that for this class of infinite-dimensional systems the first steps of the development carry through with little change. Following the exposition in Francis [75] we consider the following standard H^∞-problem.

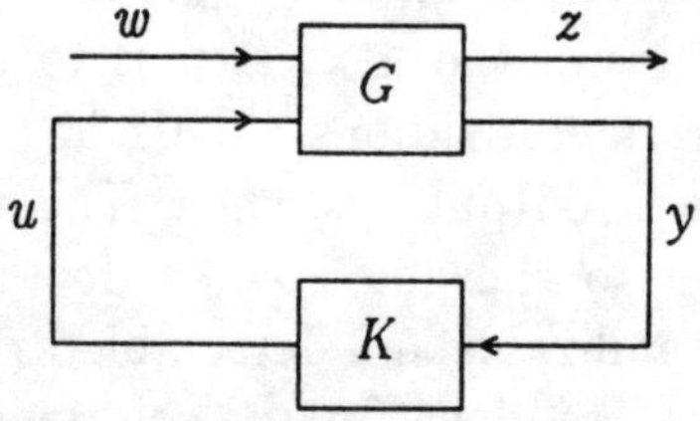

FIG. 5.4. *The standard block diagram for H^∞-problems.*

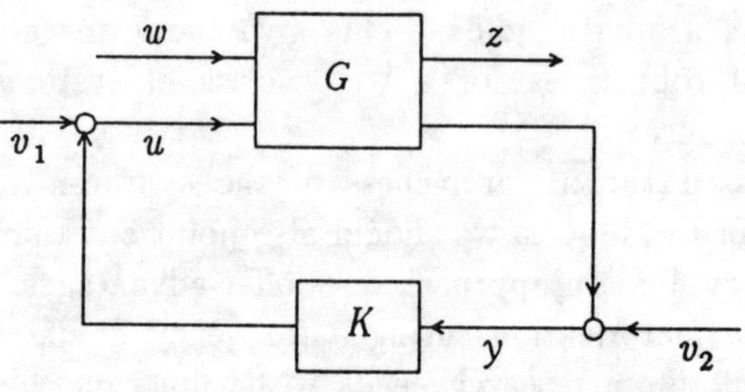

FIG. 5.5. *Diagram for the definition of stability for H^∞-problems.*

DEFINITION 5.7.1 (THE STANDARD H^∞-PROBLEM). Suppose that (C_i, A, B_j) is a Pritchard–Salamon system for $i, j = 1, 2$ and that (C_2, A)

is exponentially detectable and (A, B_2) is exponentially stabilizable. Consider Fig. 5.4 in which the four blocks of the transfer matrix of the plant G are given by

(5.61) $$G_{ij}(s) := D_{ij} + C_i(s - A)^{-1}B_j; \qquad i, j = 1, 2,$$

where D_{ij} are $p \times m$ matrices and $D_{22} = 0$. The standard H^∞-problem is to find a $K \in \mathcal{M}(\hat{\mathcal{B}}(0))$ that minimizes the following H^∞-norm:

(5.62) $$\inf \|G_{11} + G_{12}K(I - G_{22}K)^{-1}G_{21}\|_\infty$$

over all $K \in \mathcal{M}(\hat{\mathcal{B}}(0))$ that stabilize the feedback configuration of Fig. 5.5, where $G = \begin{pmatrix} G_{11} & G_{12} \\ G_{21} & G_{22} \end{pmatrix}$, i.e., the transfer functions from

$$\begin{pmatrix} w \\ v_1 \\ v_2 \end{pmatrix} \to \begin{pmatrix} z \\ u \\ y \end{pmatrix} \quad \text{are in } \mathcal{M}(\hat{\mathcal{A}}_-(0)).$$

The optimization problem (5.62) is a complicated nonstandard one; it is nonlinear in K and K ranges over the class of feedback stabilizers which is only defined implicitly. However, using the Youla parametrization of stabilizing controllers from Theorem 5.2.2 and the coprime factorization of G from Corollary 5.3.2, one can reduce this problem to an equivalent affine minimization problem over a variable $Q \in \mathcal{M}(\hat{\mathcal{A}}_-(0))$, which is a much simpler problem. This problem is called the model matching problem.

THEOREM 5.7.1. *The standard H^∞-problem of Definition* 5.7.1 *reduces to the following model matching problem*

(5.63) $$\inf_Q \|T_1 - T_2QT_3\|_\infty,$$

where $Q \in \mathcal{M}(\hat{\mathcal{A}}_-(0))$ and $T_i \in \mathcal{M}(\hat{\mathcal{A}}_-(0))$ are given by

(5.64)

$$T_1(s) = D_{11} + [C_1 + D_{12}F; -D_{12}F]\left(s - \begin{pmatrix} A_F & -B_2F \\ 0 & A_H \end{pmatrix}\right)^{-1} \begin{bmatrix} B_1 \\ B_1 + HD_{21} \end{bmatrix},$$

(5.65) $$T_2(s) = D_{12} + (C_1 + D_{12}F)(s - A_F)^{-1}B_2,$$

(5.66) $$T_3(s) = D_{21} + C_2(s - A_H)^{-1}(B_1 + HD_{21}),$$

where F is such that $A = (A + B_2F)$ is exponentially stable, and H is such that $A_H = (A + HC_2)$ is exponentially stable.

Proof. This is now an easy consequence of our control synthesis results in §§5.2 and 5.3, which justifies the application of all steps in [75, Chaps. 2–4]. The arguments are similar to those found in [42, Lemma 4.10] and [53, Lemma 3.3].

This type of generalization was also alluded to by Vidyasagar in §8.3 in [146], but without the results of Lemmas 5.3.2 and 5.3.5, it is not possible to interpret the formulas (5.64)–(5.65) sensibly.

The difficult step is to reduce (5.63) to a Nehari or general distance problem. In the finite-dimensional case (see [75]) the step between the model matching problem and the Nehari problem is achieved via various state-space spectral factorization techniques based on [11]. However, extensions of these results to general infinite-dimensional systems are only partial (see [9], [10], [31]) and do not apply directly to the symmetric factorization which is used in Chapters 7 and 8 of [75] for the rational case. Pure frequency domain results on spectral factorization are known [15].

However, recently, a way around these obstacles has been found and a coprime factorization solution to the H^∞-control problem was obtained in [51] in terms of J-spectral factorizations. This solution is valid for the subclass of Callier–Desoer class of systems for which the impulse response is of the form $f_a(t) + f_0\delta(t)$; in other words, no feedthrough delay terms are allowed. In particular, this solution applies to the class with a state-space formulation $G(s) = D + C(sI - A)^{-1}B$, where (C, A, B) belongs to the Pritchard–Salamon class and A has at most finitely many unstable poles. The problem is that the "solution" is in terms of J-spectral factorizations, and it is desirable to translate this solution into something more tangible, such as state-space Riccati equations. So far, this has been done only for the case in which B and C are bounded operators.

Earlier solutions to the infinite-dimensional H^∞-control problem had been obtained by a direct state-space approach along the lines of the finite-dimensional approaches in [86], [64], [100], [141], [159]. The solution is in terms of two nonstandard, coupled Riccati equations. The first contribution was for the special problem of the robust stability radius in [132]–[134] and [143] (this problem is equivalent to a singular H^∞-problem with complete information). They obtained a solution for the Pritchard–Salamon class in terms of a single Riccati equation with a small ϵ-parameter. Later, the regular H^∞-optimal control problem for systems with bounded input and output operators appeared in [144], [94], and [145] and for time-varying systems in [142]; for a survey of these recent developments, see [45]. A challenging area of future research is to extend these results to distributed systems with unbounded inputs and outputs.

Direct frequency domain techniques have proved to be very successful in solving one particular H^∞-problem, namely the sensitivity problem for delay systems in [66]–[68], [73], [74]. Recently, progress has been made in more general H^∞-problems in [70]–[72], [122]–[124], and [101] (for a survey, see [123a]). While the most general class considered in [70]–[72] is much larger than the Pritchard–Salamon class, the numerical aspects, especially that of approximating the optimal controller, still need to be worked out. This is nontrivial, since the optimal controller is infinite-dimensional and often improper, and not all H^∞-problems have nice continuity properties with respect to changes in the plant (see [140], [65]). Some useful contributions in the numerical aspects of finite-dimensional controller can be found in [123], [105], [69], and [135], but many of the problems are still open.

Summarizing, H^∞-problems in infinite dimensions represents a challenging new research area and both frequency domain and state-space techniques offer promising approaches.

The state-space approach leads directly to two nonstandard coupled Riccati equations, and the resulting optimal controller is infinite-dimensional. So again, an approximation theory is required to enable the design of a finite-dimensional controller. While an approximation theory for these more general Riccati equations along the lines in [97] and [104] is certainly viable, it is interesting to note that in the case of the two special H^∞-control problems considered in §§5.5 and 5.6, a more direct approximation via the transfer function was possible. Moreover, this approximation gave bounds on the performance which could be achieved by the approximating finite-dimensional controller. It is this experience which suggests that a synthesis of time and frequency domain methods for these problems offers certain advantages. For the classes of systems that do not have a nice state-space realization (for example, the cascade of a delay with a rational transfer function), a direct frequency domain approach is the natural one to follow. Moreover, this may also be preferable in the case of systems with very unbounded B and C operators.

5.8. General Remarks on Robustness

The main applications in this article are two theories of robust controller design for a strictly proper system which is the sum of an unstable infinite-dimensional part and a stable finite-dimensional part in $H^\infty(-\epsilon)$. In order to design finite-dimensional controllers, we need to be able to approximate the stable part by a finite-dimensional system in the L_∞-norm. This is always possible for strictly proper systems with an impulse response in $\hat{\mathcal{A}}(0)$ and we have discussed various approximation techniques in §5.4. Notice that all calculations are finite-dimensional ones and that we avoided having to solve infinite-dimensional Riccati equations; they were only used at a theoretical level to prove existence. These theories apply to a wide class of infinite-dimensional systems, the main shortcoming being that the perturbations are *additive* and *unstructured* and that systems with *infinitely many unstable poles* are excluded. It is therefore appropriate to explore possibilities of ameliorating these limitations.

Interesting results on the harder problem of robust stabilization for unstructured, *multiplicative* perturbations can be found in [147] in a general ring setting and in [102] for a special class of infinite-dimensional systems with scalar inputs and outputs using an interpolation approach. However, in most applications, and especially those to flexible systems, one is more interested in *structured* perturbations and studies show that the unstructured perturbation theories lead to very conservative results [25], [20], [26]. Although one can formulate robustness under structured perturbations as an H^∞-control problem [114], its solution is very difficult even for finite-dimensional systems. A state-space formulation is very appealing intuitively, but while a satisfactory theory exists for robustness of stability under open-loop perturbations, only very partial results have been obtained for the design of robust controllers [133], [134]. Perhaps an integrated time and frequency domain approach using some of the ideas proposed in [17], [18] would be successful in infinite dimensions?

The case where a system has *infinitely many unstable poles* is extremely interesting and it has attracted much attention in the literature. While the Pritchard–Salamon class does not exclude this possibility, as soon as we require exponential detectability or stabilizability, we are reduced to the finitely many unstable poles situation (Lemma 5.3.4). Due to technical reasons, the theory for robustness under additive perturbations discussed in §5.5 cannot be extended to include systems with infinitely many unstable poles. However, the factor robustness theory of §5.6 does not have these limitations and Georgiou and Smith in [78] have extended it to the very general class, $\mathcal{M}(H^\infty/H^\infty)$, the class of transfer matrices whose components are in the quotient field of H^∞ (see also [54]). This class is as general as could be wished and allows for infinitely many unstable poles or even improper systems. Applications of this theory to specific robust controller designs are discussed in §§5.6 and 5.9. Here we take the point of view that we have a meta theory of robustness for linear infinite-dimensional systems which are input-output stabilizable. This theory applies to all our examples and in particular to Examples 5.2.6 and 5.2.7, which have infinitely many unstable poles and were not covered by our theories in §§5.5 or 5.6.

If the robustness theory of [78] applies to our Examples 5.2.6 and 5.2.7, then how do we explain the anomaly that these systems are *destabilized by arbitrarily small delays*? Are these systems robustly stable or not? Recently, there has been a spate of papers proving that certain clever velocity feedback stabilizing schemes for various p.d.e. models of undamped flexible beams are embarrassingly nonrobust; they can be destabilized by arbitrary small delays in the feedback loop, e.g., Examples 5.2.6, 5.2.7, [58], [59]. This is unsatisfactory since most control laws are implemented digitally and small delays are inevitable. Of course, this phenomenon is not new in engineering circles and it has been known for a long time that even finite-dimensional improper systems can be destabilized by arbitrarily small delays [154]; and this has been shown to be true for several infinite-dimensional systems using a direct, simple frequency domain analysis, [8], [13], [14], [27]. At present this phenomenon does not appear to be well understood in p.d.e. circles, but I think that a simple explanation can be found using a frequency domain analysis analogous to the recent paper [79] on w-stability finite-dimensional systems. There it is noted that the factor robustness discussed in §5.6 does not necessarily imply robustness against certain kinds of perturbations, in particular, arbitrarily small delays; these are perturbations which induce a large value in the T-gap metric. So it is necessary to impose extra conditions on the systems to ensure this type of robustness as well. They call this *w-stability*, and they prove that the finite-dimensional feedback system (G, K) will be w-stable if and only if (G, K) is input-output stable and $\|G(\infty)\| \, \|K(\infty)\| < 1$. In particular, strictly proper systems have this nice property.

Such a result for infinite-dimensional systems would be invaluable for understanding robustness to small delays. Here the situation is more complicated and it is necessary to be careful about the definition of proper. Recall that in Definition 5.2.3 we defined G to be *proper* if $G(s)$ has a limit as $|s| \to \infty$ in $\operatorname{Re} s > 0$ and G is *strictly proper* if this limit is the zero matrix. Moreover, G was defined to be *improper* if there exist s_n in $\operatorname{Re} s > 0$ such that $|s_n| \to \infty$ and $\|G(s_n)\| \to \infty$. Notice that G need not have a limit at infinity, but if it does we denote this by $G(\infty)$. While it is difficult to obtain necessary and sufficient conditions for robustness to arbitrarily small delays, a careful analysis of the proofs in [79] shows that the following partial results hold for G, K belonging to $\mathcal{M}(H^\infty/H^\infty)$.

LEMMA 5.8.1 ([110]). *If G and $K \in \mathcal{M}(H^\infty/H^\infty)$ are proper, then (G, K) will be w-stable if and only if*

(i) *(G, K) is input-output stable;*

(ii) $\|G(\infty)\| \, \|K(\infty)\| < 1$.

These partial results already give us sufficient conditions for w-stability of proper Callier–Desoer systems. In particular, the robust controller design of §5.5 assumes that the plant is strictly proper and from (5.38) we see that the controllers are always proper. Lemma 5.8.1 then shows that the robust controllers of §5.5 will also be w-stable. If one examines carefully the full parametrization of the robust controllers of §5.6 one sees that not all the (possibly infinite-dimensional) controllers will be proper, but that by choosing the free parameter in Theorem 5.4.2 to be proper we do obtain proper controllers. Our plant is strictly proper and so again Lemma 5.8.1 implies w-stability. Similarly, if we design finite-dimensional controllers by the approximation methods discussed in §§5.5 and 5.6 we also have w-stability. Summarizing, the robust controller designs we discussed in §§5.5 and 5.6 produce w-stable controllers and will not be destabilized by arbitrarily small delays. For scalar Callier–Desoer systems this conclusion also follows from earlier results in [8], which considered destabilization by small delays for scalar infinite-dimensional systems.

Let us now examine Examples 5.2.6 and 5.2.7, which are stabilized by a direct feedthrough controller of the type $u = -ky$ with $k > 0$. Although the controller is proper, a careful examination of the transfer functions shows us that they are improper and so Lemma 5.8.1 does not apply. Moreover, the proof of Theorem 1 in [79] does not generalize directly to show that if either G or K is improper, then (G, K) is not w-stable. However, we already know from [58], [13], [14], and [27] that these systems are destabilized by arbitrary small delays and so they are not w-stable. In the light of the insights gained from reading the papers [8], [13], [79], it is tempting to propose the following conjecture.

CONJECTURE 5.8.1. *If G is in $\mathcal{M}(H^\infty/H^\infty)$ and is improper, then G is not w-stabilizable.*

In view of the many p.d.e. models of undamped flexible structures that have an improper transfer function, the proof or disproof of this conjecture would be an important contribution to our understanding of w-stability. It is interesting to note that improper systems turn out to be "ill-posed" according to the design criteria formulated in [91]. In fact, they claim that the only frequency domain models which produce good control laws in an engineering sense are those belonging to the Callier–Desoer class (see Theorem 5.2.3). This class certainly has nice robustness properties with respect to additive, and factor perturbations as we saw in §§5.5 and 5.6 and w-robustness is covered by Lemma 5.8.1. However, there are still many challenging unanswered questions concerning different types of robustness for improper systems like Examples 5.2.6 and 5.2.7 and the recent paper [79] suggests that frequency domain techniques might prove fruitful here.

5.9. Extensions and Future Directions

In §§5.2–5.7 we have discussed an integrated system theoretic approach using time and frequency domain methods applied to the Callier–Desoer class of transfer functions and the Pritchard–Salamon state-space class. These are special classes of infinite-dimensional systems; the Pritchard–Salamon class has an integrable impulse response and the Callier–Desoer class can have only finitely many unstable poles, and both classes have proper transfer functions. It is interesting to ask if this synthesis is possible for a larger class of systems.

An obvious extension of the Callier–Desoer class of transfer functions is the quotient field of H^∞, H^∞/H^∞, which was introduced in §5.8. If we restrict ourselves to input-output stabilizable systems, then all the algebraic properties of the Callier–Desoer class are retained, including the key one of coprime factorization. Consequently, the class of stabilizable systems in $\mathcal{M}(H^\infty/H^\infty)$ can serve as a convenient meta class of transfer matrices, which even includes improper systems. However, sometimes it is convenient to study special classes of systems with special properties (see [148] and [146, Chap. 8] for a discussion of this point). In this direction we mention the *Logemann* class [107], which comprises transfer matrices that are the sum of a finite-dimensional unstable part and a stable part in $\mathcal{M}(H^\infty(-\epsilon))$. This means that this class has only finitely many unstable poles, but it does admit less smooth transfer functions than the Callier–Desoer class. Its structure was exploited in [107] to prove the existence of finite-dimensional compensators for retarded and neutral systems, in [108] to prove the equivalence of input-output stability and exponential stability for retarded and neutral systems and in [112], [113] to extend classical frequency domain results on P.I. controllers for minimum-phase systems to an infinite-dimensional setting. As mentioned at the end of §5.5, the theory of robust stabilization under additive perturbations extends naturally to this class [109]. Moreover, the recent results from [43] show that if the transfer function has a well-posed realization in the sense of Salamon [137], [138], [55] which is also exponentially stabilizable and detectable, then we obtain exponential stability. Other special classes have been proposed which include proper, nonrational transfer functions which have infinitely many poles on the imaginary axis: these are the *strictly noncyclic* (roomy) class considered in [7], [77], [6] and the *pseudorational* class considered in [155]–[157]. These classes also form algebras and most of the concepts defined in §5.2 for the Callier–Desoer class can be extended to these classes, including the existence of coprime factorizations. The pseudorational class is appropriate for retarded and neutral systems and has been exploited to prove the equivalence of internal and external stability in [157] and [156]. Although several delay and p.d.e. examples are known to belong to the strictly noncyclic class, which state-space systems it covers is not yet well understood; you first need to obtain the transfer function and directly verify if it belongs to the transfer function class. Perhaps this is the reason that the obvious potential of the theory of the strictly noncyclic class (which was known in the seventies and clearly documented in [7] and [77]) has never been fully exploited. For the precise definitions, we refer to [77], [7], [157]; the important features are that they are classes of proper transfer functions which may have infinitely many, very nasty singularities on the imaginary axis. They have a nonzero intersection with the Logemann class and with each other, but none seems to be contained in any other. The strictly noncyclic class has most of the nice algebraic properties such as coprime factorizations which are needed to effect the controller synthesis advocated by Vidyasagar [146], as has been pointed out by Baras in [7] and [6]. So one has a Youla parametrization of all stabilizing controllers as in Theorem 5.2.2 and conditions are known for the existence of dynamic controllers [6]. Recently, progress has been made with algorithms for computing coprime factors [16] and so all the necessary ingredients for control synthesis are available. Of course, the resulting controllers will be infinite-dimensional, which some may find a disadvantage [91]. However, Baras is a great believer in the feasibility of infinite-dimensional controllers, the classic example being that of a pure delay. Be this as it may, the classes of strictly noncyclic and pseudorational transfer functions have some very attractive features from the point of view of control synthesis and deserve more attention in the future.

The logical extension of the Pritchard–Salamon class of state-space systems is the *Salamon–Weiss* class introduced in [137], [138] to describe most well-posed delay and p.d.e. systems, (including Examples 5.2.3, 5.2.5, and 5.2.6). To define this class we suppose that there exist three Hilbert spaces with continuous and dense inclusions: $W \hookrightarrow H \hookrightarrow V$ and that A generates a consistent C_0-semigroup on all three spaces. Then using the interpretation of the Salamon–Weiss class from [40], we say that (C, A, B) belongs to the Salamon–Weiss class if $C \in \mathcal{L}(W, \mathbb{R}^p)$ and induces a bounded observability map on H, $B \in \mathcal{L}(\mathbb{R}^m, V)$ and induces a bounded reachability map on H and the *generalized transfer function* $G^\mu(s) := D + (\mu - s)C(s-A)^{-1}(\mu - A)^{-1}B$ is in $H^\infty(\alpha)$ for $\mu \in \rho(A)$ and some $\alpha > 0$. In other words, $G^\mu(\cdot)$ is analytic and bounded in some open right half plane $\operatorname{Re} s > \alpha$; $G^\mu(\cdot) \in \mathcal{M}(H^\infty(\alpha))$. Many properties of this class have been studied in [137], [138], [40], [151]–[153], [55] and a recent survey may be found in [43]. However, the known theory for this class is not as complete as for the Pritchard–Salamon class. Partial results are available on the existence of solutions of the Nehari problem [1], [17], on the existence of balanced realizations [121], and some approximation results can be found in [85] and [120]. However, there remain many open problems to be solved before an elegant synthesis for the Salamon–Weiss class can be developed; in particular, sharper perturbation results, explicit representations of coprime factorizations and a unified operator theoretic approach for algebraic Riccati equations of this class are needed.

Another way to synthesize state-space and frequency domain approaches can be illustrated with reference to the problem of robust stabilization with respect to factor perturbations. Although there exists a complete frequency domain solution given in [78], to obtain an answer in state-space terms one needs to find normalized coprime factorizations and to solve a Nehari problem as in §5.6. In [54] these steps have been successfully done for a special class of systems with infinitely many unstable poles and so one might try doing the same for other special classes of infinite-dimensional systems for which the relevant Riccati equations have well-defined solutions (see the recent survey [104]). Of course, this has the disadvantage that these technical steps would have to be repeated for each new example which is rather tedious. A unified approach for the general Salamon–Weiss class would be more efficient and elegant.

In §5.7, we already reported on pure state-space approaches to the general H^∞-optimal control problem. So far, solutions have been obtained for the limited class of distributed systems with bounded B and C operators. A challenging area of future research is to obtain analogous solutions for a wider class of distributed systems, as has been done for the LQ Riccati equation (see [104]). This will be much harder than the LQ-case, as the more general Riccati equations include BKC-perturbation terms which are technically more difficult to handle. Another challenging area of future research is the design of finite-dimensional H^∞-controllers; as is the case for LQ-controllers the optimal H^∞-controllers are infinite-dimensional. One approach is to develop an approximation theory for these H^∞-Riccati equations along the lines described in [104] for the special LQ-case, and then to use this to develop a compensator design as in [97]. In the opinion of this author, it should prove more efficient to use a synthesis of state-space and frequency domain approaches, as was done for the two special H^∞-control problems in §§5.5 and 5.6. They were numerically simpler and they gave valuable information on the achievable performance.

Although the central theme of this article has been the synthesis of frequency and time-domain techniques, it is clear from the general remarks on robustness in §5.8 that frequency domain techniques offer a powerful approach for analyzing systems. As well as our remarks on the Logemann, strictly noncyclic and pseudorational classes, we have already mentioned very successful, purely frequency domain techniques to robustness at the end of §§5.6 and 5.8 and to H^∞-control at the end of §5.7. Noteworthy is that a direct frequency domain approach can be used to find coprime factors [16], thus avoiding the intermediate step via Riccati equations which we used in our analysis of the Pritchard–Salamon class. For systems for which Riccati equations cannot be posed or for which these involve delicate technicalities, this could prove a better approach. While it is outside the scope of this chapter to survey frequency domain techniques, it is worthwhile mentioning the excellent account in [91] of topics in analytic function theory, such as interpolation, factorization, H^∞-theory and approximation, which can be applied to many control problems that have been formulated in a frequency domain setting. Earlier steps in this direction can be found in [158] and [30] and more recently there have been many examples of frequency domain solutions of control problems for infinite-dimensional systems, for example, [6], [13], [66]–[68], [70]–[74], [78]–[80], [101], [105], [122]–[124], [135], [160], [165], and [99]. (This list is meant to be representative rather than comprehensive.) In the future we can expect more results using the powerful techniques from classical and from modern operator theory.

5.10. Conclusions

As stated in the introduction, the focus of this paper has been on the Callier–Desoer transfer function class and the Pritchard–Salamon state-space class and their relationships. The advantage is that we have identified useful classes of linear infinite-dimensional systems which are amenable to the systems theoretic approach to control synthesis that has been so successful in finite dimensions [146], [75], [62], [81], [82], [114]. The most striking feature of these recent developments is the increasing integration of state-space and frequency domain techniques [17], [86], [64], [100], and [159]; both techniques are used in a complementary way, and we have seen that this can be done in infinite dimensions as well.

The applications of this synthesis approach were to solve two different robustness problems. In §5.5 a complete solution for robust stabilization with respect to additive perturbations was solved for strictly proper systems in the Callier–Desoer class. In §5.6 the problem of robust stabilization with respect to factor perturbations was solved for exponentially stabilizable and detectable Pritchard–Salamon systems. Explicit solutions were given in terms of Riccati equations.

Combining the theoretical robustness theories of §§5.5 and 5.6 with the approximation results of §5.4 led to practical control designs for an even larger class of systems: those which are the sum of an unstable finite-dimensional part and a stable part in $H^\infty(-\epsilon)$ which is approximable in the L_∞-norm. The control designs were essentially finite-dimensional designs which could be performed on PCMATLAB [129]. The techniques were illustrated by a flexible beam model, but are also applicable to delay systems. Notice that we do not have to solve the infinite-dimensional Riccati equations to design robust controllers with this approach.

In §5.7 more general H^∞-control problems incorporating other performance criteria were formulated for the Pritchard–Salamon class. Recent results on both state-space and frequency domain solutions were discussed and further areas of research were suggested.

In §5.8 some recent insights into robustness were discussed, in particular, the phenomenon of destabilization by arbitrarily small delays. This turns out to be a special case of a more general phenomenon and has been analyzed under the name w-stability using a frequency domain approach.

Finally, some ideas on extending the synthesis of time and frequency approaches to larger classes of systems were discussed in §5.9, and some related work on frequency domain approaches was mentioned.

In all sections various open problems were discussed, not only state-space problems and frequency domain problems, but especially questions concerning the relationships between the two approaches: the synthesis theme. It is hoped that this article will help to stimulate more research into infinite-dimensional systems using both time and finite-dimensional techniques in an integrated system theoretic approach.

5.11. Glossary of Notation

Symbol	*Location in the text*
t denotes the matrix transpose	
$\mathcal{A}(\mu)$, $\hat{\mathcal{A}}(\mu)$, $\hat{\mathcal{A}}_-(\mu)$	Definition 5.2.1
$\hat{\mathcal{B}}(\mu)$, $\hat{\mathcal{A}}_-^\infty(\mu)$, $\mathcal{M}(\hat{\mathcal{B}}(\mu))$, $\mathcal{M}(\hat{\mathcal{A}}_-(\mu))$	Definition 5.2.2
$\mathbb{C}_\mu^+ = \{s : \operatorname{Re} s \geq \mu\}$, $\mathbb{C}_\mu^- = \{s : \operatorname{Re} s < \mu\}$	Definition 5.2.2
$\mathcal{R}^\infty(\mu)$	Theorem 5.2.1
H^∞ the space of matrix valued functions of a complex variable which are analytic and bounded in $\operatorname{Re} s > 0$	after Definition 5.2.1

$H^\infty(-\epsilon)$	the space of matrix-valued functions of a complex variable which are analytic and bounded in Re $s > -\epsilon$	in §5.8
H^∞_-	the space of matrix-valued functions of complex variable which are analytic and bounded in Re $s < 0$	Theorem 5.4.1
$\mathcal{L}(X, Y)$	the space of bounded linear operators from X to Y	Definition 5.3.1
$C(0, t_1; W)$	the space of continuous functions: $[0, t_1] \to W$	above (5.14)
$\mathcal{B}$	the reachability map	(5.15)
$\mathcal{C}$	the observability map	(5.16)
Γ	the Hankel operator	(5.17)
σ_i	Hankel singular values	below (5.17)
P	solution of the filter algebraic Riccati equation	(5.22)
Q	solution of the control algebraic Riccati equation	(5.20)
P_0	controllability gramian	Theorem 5.4.2
Q_0	observability gramian	
$\mathcal{F}_L$	linear fractional transformation	(5.49)
$\sigma_{\min}(G)$	the smallest Hankel singular value of the finite-dimensional system G.	(5.37)

Acknowledgments.
I would like to express my thanks to my colleagues John Baras, Denis Bernstein, Jan Bontsema, Frank Callier, Bruce Francis, Keith Glover, Rien Kaashoek, Pramod Khargonekar, Hartmut Logemann, André Ran, Hans Schumacher, Allen Tannenbaum, Stuart Townley, Mathukumalli Vidyasagar, Jan Willems, and Hans Zwart who took the time to read the preliminary draft and to offer suggestions for its improvement. I am also grateful to the *neutral reviewer* who made some positive suggestions for improving this manuscript, in particular, the inclusion of the beam example which was the main object of study in Jan Bontsema's thesis [20]. The *passionate reviewer* convinced

me that writing this paper was really worthwhile and gave me the necessary incentive to update the preliminary years-old version. Finally, thanks are due to Janieta Schlukebir who has so willingly typed so many revisions of this manuscript for me.

This research was partly supported by the Netherlands Foundation for the Technical Sciences (STW) project GWI 33.0533 and the European Economic Community (EEC) contract ST2J-0066-3-NL (CD).

References.

[1] V. M. Adamjan, D. Z. Arov, and M. G. Krein, *Analytic properties of Schmidt pairs for a Hankel operator and the generalized Schur–Takagi problem*, Math. USSR-Sb., 15 (1971), pp. 31–73.

[2] A. V. Balakrishnan, *Boundary Control of Parabolic Equations:* LQR-*theory*, Proc. Conference Theory of Non Linear Equations, Nonlinear Operators, Proc. 5th Internat. Summer School, Akademic-Verlag, Berlin, 1978.

[3] M. Balas, *Towards a (more) practical control theory for distributed parameter systems*, in Control and Dynamic Systems: Advances in Theory and Applications, Vol. 18, C. T. Leondes, ed., Academic Press, New York, 1980.

[4] J. A. Ball and J. W. Helton, *A Beurling–Lax theorem for the Lie group $U(m,n)$ which contains most classical interpolation theory*, J. Operator Theory, 8 (1983), pp. 107–142.

[5] S. P. Banks, *State-Space and Frequency-Domain Methods in the Control of Distributed Parameter Systems*, P. Peregrinus, London, 1983.

[6] J. Baras, *Frequency domain design of linear distributed systems*, Proc. IEEE Conference on Decision and Control, 1980.

[7] J. Baras and P. Dewilde, *Invariant subspace methods in linear multivariable-distributed systems and lumped distributed network synthesis*, Proc. IEEE, 64 (1976), pp. 160–178.

[8] J. F. Barman, F. M. Callier, and C. A. Desoer, *L^2-Stability and L^2-instability of linear time-invariant distributed feedback systems perturbed by a small delay in the loop*, IEEE Trans. Automat. Control, 18 (1973), pp. 479–484.

[9] H. Bart, I. Gohberg, and M. A. Kaashoek, *Wiener–Hopf factorization, inverse Fourier transform and exponentially dichotomous operators*, J. Funct. Anal., 68 (1986), pp. 1–42.

[10] ——, *Wiener–Hopf equations with symbols analytic in a strip*, Oper. Theory: Adv. Appl. 21, Birkhäuser Verlag, Basel, 1986, pp. 39–74.

[11] H. Bart, I. Gohberg, M. A. Kaashoek, and P. Van Dooren, *Factorizations of transfer functions*, SIAM J. Control Optim., 18 (1980), pp. 675–696.

[12] R. Bellman and K. L. Cooke, *Differential-Difference Equations*, Academic Press, New York, 1963.

[13] W. H. Bennet, H. G. Kwatny, and J. S. Baras, *Frequency domain design of robust controllers for large space structures*, Report SEI TR-86-11, Systems Engineering Inc., Greenbelt, MD, 1986.

[14] ——, *Robustness issues in boundary feedback of flexible structures*, Proc. ACC, 1987.

[15] W. H. Bennet and Ing Yan, *A computer algorithm for causal spectral factorization*, Proc. NAECON 88, Dayton, OH, May 1988.

[16] C. A. Berenstein and D. C. Struppa, *On explicit solutions to the Bezout equation*, Systems Control Lett., 4 (1984), pp. 33–39.

[17] D. S. Bernstein and W. M. Haddad, LQG *Control with an H^∞-performance bound: a Riccati equation approach*, IEEE Trans. Automat. Control, 34 (1989), pp. 293–305.

[18] ——, *The optimal projection equation with Petersen–Hollot bounds: robust stability and performance via fixed-order dynamic compensation for systems with structured real-valued parameter uncertainty*, IEEE Trans. Automat. Control, 33 (1988), pp. 578–582.

[19] ——, *The optimal projection equations for finite-dimensional fixed-order dynamic compensation of infinite-dimensional systems*, SIAM J. Control Optim., 24 (1986), pp. 122–151.

[20] J. Bontsema, *Dynamic stabilization of large flexible space structures*, Ph.D. thesis, June 1989, University of Groningen, Groningen, the Netherlands.

[21] J. Bontsema and R. F. Curtain, *Perturbation properties of a class of infinite-dimensional systems with unbounded control and observation*, J. Math. Control Inform., 5 (1988), pp. 333–352.

[22] ——, *A note on spillover and robustness of flexible systems*, IEEE Trans. Automat. Control, 33 (1988), pp. 567–569.

[23] ——, *Comparison of some controllers on a flexible beam*, Proc. 25th IEEE Conf. on Decision and Control, 1986, Athens, IEEE Control Systems Soc., New York, 1986, pp. 165–166.

[24] ——, *Robust stabilization of a flexible beam using a normalized coprime factorization approach*, Control of Uncertain Systems, D. Hinrichsen and B. Mårtensson, eds., Proceedings of an International Workshop, Bremen, Germany, Birkhäuser, Boston, MA, 1990, pp. 1–18.

[25] J. Bontsema, R. F. Curtain, and J. M. Schumacher, *Robust control of flexible structures: a case study*, Automatica–J. IFAC, 24 (1988), pp. 177–186.

[26] J. Bontsema and T. Van De Vaart, *Comparison of robustness measures for a flexible structure*, M. A. Kaashoek, J. H. van Schuppen, and A. C. M. Ran., eds., Proc. Internat. Sympos. MTNS, Vol. II, Amsterdam, Birkhäuser, Boston, MA, 1990.

[27] J. Bontsema and S. A. DeVries, *Robustness of flexible systems against small time delays*, 27th IEEE Conference on Decision and Control, Los Angeles, CA, 1988.

[28] F. M. Callier and C. A. Desoer, *An algebra of transfer functions for distributed linear time-invariant systems*, IEEE Trans. Circuits and Systems, 25 (1978), pp. 651–663. (Corrections: 26 (1979), p. 320.)

[29] ——, *Simplifications and new connections on an algebra of transfer functions of distributed linear time-invariant systems*, IEEE Trans. Circuits and Systems, 27 (1980), pp. 320–323.

[30] ——, *Stabilization, tracking and disturbance rejection in multivariable convolution systems*, Ann. Soc. Sci. Bruxelles, 94 (1980), pp. 7–51.

[31] F. M. Callier and J. Winkin, *On spectral factorization and LQ-optimal regulation for multivariable distributed systems*, Internat. J. Control, 52 (1990), pp. 55–75.

[32] G. Chen, M. C. Delfour, A. M. Krall, and G. Payre, *Modelling, stabilization and control of serially connected beams*, SIAM J. Control Optim., 25 (1987), pp. 526–546.

[33] M. Chen and C. A. Desoer, *Necessary and sufficient conditions for robust stability of distributed feedback systems*, Internat. J. Control, 35 (1982), pp. 255–267.

[34] R. F. Coifman and R. Rochberg, *Representation theorems for holomorphic and harmonic functions in L^p*, Asterisque, 77 (1980), pp. 11–66.

[35] R. F. Curtain and A. J. Pritchard, *Infinite dimensional linear systems theory*, Lecture Notes in Control and Inform. Sci. 8, Springer-Verlag, Berlin, 1978.

[36] R. F. Curtain, *Spectral systems*, Internat. J. Control, 39 (1984), pp. 657–666.

[37] R. F. Curtain, *Sufficient conditions for infinite-rank Hankel operators to be nuclear*, IMA J. Math. Control Inform., 2 (1985), pp. 171–182.

[38] ——, ed., *Modelling, robustness and sensitivity reduction in control systems*, NATO Adv. Sci. Inst. Ser. F, Comput. Systems Sci., 34, Springer-Verlag, Berlin, New York, 1986.

[39] ——, *Equivalence of input-output stability and exponential stability for infinite-dimensional systems*, J. Math. Systems Theory, 21 (1988), pp. 19–48.

[40] ——, *Well-posedness of infinite-dimensional linear systems in time and frequency domain*, 1988, Report TW-287, Mathematics Institute, University of Groningen, Groningen, the Netherlands.

[41] ——, *L_∞-approximations of complex functions and robust controllers for large flexible space structures*, Nieuw Arch. Wisk., 6 (1988), pp. 7–22.

[42] ——, *Robust stabilizability of normalized coprime factors; the infinite-dimensional case*, Internat. J. Control, 51 (1990), pp. 1173–1190.

[43] ——, *Representation of infinite-dimensional linear systems*, pp. 101–128, Three Decades of Mathematical System Theory, A Collection of Surveys at the Occasion of the Fiftieth Birthday of Jan C. Willems, H. Nijmeijer and J. M. Schumacher, eds., Lecture Notes in Control and Inform. Sci. 135, Springer-Verlag, Berlin, New York, 1989.

[44] ——, *Equivalence of input-output stability and exponential stability*, Systems Control Lett., 12 (1989), pp. 235–239.

[45] ——, *State-space approaches to H^∞-control for infinite-dimensional linear systems*, Trans. Inst. Math. Control, 13 (1991), pp. 253–261.

[46] R. F. Curtain, A. Bensoussan, and J. L. Lions, eds., *Analysis and optimization of systems: state and frequency domain approaches to infinite-dimensional systems*, Proc. 10th Internat. Conf., Sophia-Antipolis, June 9–12, 1992, Lecture Notes in Control and Inform. Sci., Springer-Verlag, Berlin, New York, to appear.

[47] R. F. Curtain and K. Glover, *Robust stabilization of infinite-dimensional systems by finite dimensional controllers: derivations and examples*, Modelling Identification and Robust Control, Proc. MTNS 1985, Sweden, C. I. Byrnes and A. Lindquist, eds., Elsevier Science Publishers, North Holland, 1986, pp. 113–128.

[48] ——, *Balanced realisations for infinite dimensional systems*, Operator Theory and Systems, Proc. Workshop Amsterdam, June 4–7, 1985, H. Bart, I. Gohberg, and M. A. Kaashoek, eds., Birkhäuser, Basel, 1986, pp. 87–104.

[49] ——, *Controller design for distributed systems based on Hankel-norm approximation*, IEEE Trans. Automat. Control, 31 (1986), pp. 173–176.

[50] ——, *Robust stabilization of infinite dimensional systems by finite dimensional controllers*, Systems Control Lett., 7 (1986), pp. 41–47.

[51] R. F. Curtain and M. Green, *H^∞-controller synthesis for infinite-dimensional systems by J-lossless coprime factorization*, manuscript.

[52] R. F. Curtain, H. Logemann, S. Townley, and H. Zwart, *Well-posedness, stabilizability and admissibility for Pritchard–Salamon systems*, Report 260, Institut für Dynamische Systeme, Universität Bremen, Bremen, Germany, 1992.

[53] R. F. Curtain and A. Ran, *Explicit formulas for Hankel norm approximations of infinite-dimensional systems*, J. Integral Equations and Operator Theory, 13 (1989), pp. 455–469.

[54] R. F. Curtain and B. A. M. Van Keulen, *Robust control with respect to coprime factors of infinite-dimensional positive real systems*, Proc. of 29th IEEE Conference on Decision and Control, 6, IEEE Control Systems Society, New York, 1990, pp. 2865–2867.

[55] R. F. Curtain and G. Weiss, *Well-posedness of triples of operators (In the Sense of Linear Systems Theory)*, Proc. 4th Internat. Conference on Control of Distributed Parameter Systems, Vorau, July 1988, in Control and Estimation of Distributed Parameter Systems, Internat. Ser. Numer. Math. 91, Birkhäuser Verlag, Basel, 1989, pp. 41–59.

[56] R. F. Curtain and H. J. Zwart, *Spectral realisations for delay systems*, Distributed Parameter Systems, Proc. 3rd Intern. Conf. Vorau 1986, F. Kappel, K. Kunisch and W. Schappacher, eds., Lecture Notes in Control and Inform. Sci. 102, 1987.

[57] ——, *L_∞-approximations of nonrational transfer functions: an example*, Proc. of the 25th IEEE Conference on Decision and Control, Dec. 10–12, 1986, Athens, Greece, IEEE Control Systems Society, New York, 1986.

[58] R. Datko, J. Lagnese and M. P. Polis, *An example on the effect of time delays in boundary feedback stabilization of wave equations*, SIAM J. Control Optim., 24 (1986), pp. 152–156.

[59] M. C. Delfour and M. P. Polis, *On issues related to stabilization of large flexible structures*, Chapter 4, this book.

[60] W. Desch and W. Schappacher, *Spectral properties of finite-dimensional perturbed linear semigroups*, J. Differential Equations, 59 (1985), pp. 80–102.

[61] C. A. Desoer, R. W. Liu, J. Murray, and R. Saeks, *Feedback system design: the fractional representation approach to analysis and synthesis*, IEEE Trans. Automat. Control, 25 (1980), pp. 399–412.

[62] J. C. Doyle, *Lecture Notes in Advances in Multivariable Control*, Office of Naval Research/Honeywell Workshop, Minneapolis, MN, 1984.

[63] ——, *Guaranteed margins for* LQG *regulators*, IEEE Trans. Automat. Control, 23 (1978), pp. 756–757.

[64] J. Doyle, K. Glover, P. Khargonekar, and B. Francis, *State-space solutions to standard H^2 and H^∞ control problems*, IEEE Trans. Automat. Control, 34 (1989), pp. 831–847.

[65] M. J. Engelhart and M. C. Smith, *A four-block problem for H^∞-design: properties and applications*, 1990, manuscript.

[66] D. S. Flamm, *Control of delay systems for minimax sensitivity*, Ph.D. thesis, LIDS-TH-1560, Massachusetts Institute of Technology, Cambridge, MA, June 1986.

[67] D. S. Flamm and S. K. Mitter, *H^∞-sensitivity for delay systems: Part* I, Systems Control Lett. 9 (1987), pp. 17–24.

[68] ——, *H^∞-sensitivity for delay systems: Part* II, Report No. LIDS-P-1647, Lab: IDS, Massachusetts Institute of Technology, Cambridge, MA, 1987.

[69] D. S. Flamm, H. Yang, Q. Ren, and K. Klipec, *Numerical computation of inner factors for distributed parameter systems*, 1991, manuscript.

[70] C. Foias and A. Tannenbaum, *On the Nehari problem for a certain class of L_∞-functions appearing in control theory*, J. Funct. Anal., 74 (1987), pp. 146–159.

[71] ——, *On the four block problem* I, Oper. Theory: Adv. Appl. 32, Birkhäuser Verlag, Basel, 1988, pp. 93–112.

[72] ——, *On the four block problem* II, Integral Equations Operator Theory, 11 (1988), pp. 726–767.

[73] C. Foias, A. Tannenbaum, and G. Zames, *Weighted sensitivity minimization for delay systems*, IEEE Trans. Automat. Control, 31 (1986), pp. 763–766.

[74] ——, *On the H^∞-optimal sensitivity problem for systems with delays*, SIAM J. Control Optim., 25 (1987), pp. 686–705.

[75] B. A. Francis, *A course in H^∞-control theory*, Lecture Notes in Control and Inform. Sci. 88, Springer-Verlag, Berlin, New York, 1987.

[76] B. A. Francis and J. C. Doyle, *Linear control theory with an H^∞-optimality criterion*, SIAM J. Control Optim., 25 (1987), pp. 815–844.

[77] P. Fuhrman, *Linear Systems and Operators in Hilbert Space*, McGraw-Hill, New York, 1981.

[78] T. T. Georgiou and M. C. Smith, *Optimal robustness in the gap metric*, IEEE Trans. Automat. Control, 35 (1990), pp. 673–686.

[79] ——, *W-stability of feedback systems*, Systems Control Lett., 13 (1989), pp. 271–277.

[80] ——, *Robust stabilization in the gap metric: controller design for distributed plants*, IEEE Trans. Automat. Control, to appear.

[81] K. Glover, *All optimal Hankel norm approximations of linear multivariable systems and their L_∞-error bounds*, Internat. J. Control, 39 (1984), pp. 1115–1193.

[82] ——, *Robust stabilization of multivariable linear systems: Relations to approximation*, Internat. J. Control, 43 (1986), pp. 741–766.

[83] ——, Private communication.

[84] ——, *A state-space approach to H^∞-optimal control*, Three Decades of Mathematical System Theory, A Collection of Surveys at the Occasion of the Fiftieth Birthday of Jan C. Willems, H. Nijmeijer and J. M. Schumacher, eds., Lecture Notes in Control and Inform. Sci. 135, Springer-Verlag, Berlin, New York, 1989, pp. 179–218.

[85] K. Glover, R. F. Curtain, and J. R. Partington, *Realisation and approximation of linear infinite dimensional systems with error bounds*, SIAM J. Control Optim., 26 (1988), pp. 899–918.

[86] K. Glover and J. Doyle, *State-space formulae for all stabilizing controllers that satisfy an H^∞-norm bound and relations to risk sensitivity*, Systems Control Lett., 11 (1988), pp. 167–172.

[87] K. Glover and D. McFarlane, *Robust stabilization of normalized coprime factor plant descriptions with H^∞-bounded uncertainty*, IEEE Trans. Automat. Control, 34 (1989), pp. 821–830.

[88] K. Glover, J. Lam, and J. R. Partington, *Balanced realisation and Hankel-norm approximation of systems involving delays*, Proc. IEEE Conf. on Decision and Control, Athens, Dec., 1986, pp. 1810–1815.

[89] ——, *Rational approximation of a class of infinite dimensional systems*, I: *Singular values of Hankel operators*, Math. Control Signals Systems, 3 (1990), pp. 325–344. II: *Optimal convergence rates of L_∞ approximants*, Math. Control Signals Systems, 4 (1991), pp. 233–246.

[90] G. Gu, P. P. Khargonekar, and E. B. Lee, *Approximation of infinite-dimensional systems*, IEEE Trans. Automat. Control, 34 (1989), pp. 610–618.

[91] A. J. Helmicki, C. A. Jacobson, and C. N. Nett, *Practical considerations in the use of distributed parameter systems in modeling: A control point of view*, IEEE Trans., 36 (1991), pp. 1053–1057.

[92] J. W. Helton, *Operator Theory,, Analytic Functions, Matrices and Electrical Engineering*, Regional Conf. Ser. in Math., 68, American Mathematical Society, Providence, RI, 1987.

[93] L. F. Ho and D. L. Russel, *Admissible input elements in Hilbert space and a Carleson measure criterion*, SIAM J. Control Optim., 21 (1983), pp. 614–640.

[94] A. Ichikawa, *Quadratic control of evolution equations with delays in control*, SIAM J. Control Optim., 20 (1982), pp. 648–668.

[95] ——, *Differential Games and H^∞-problems*, Proc. Math. Theory of Networks and Systems, 1991, Tokyo, Japan, to appear.

[96] Y. Inouye, *Parametrization of compensators for linear systems with transfer functions of bounded type*, Tech. Report 88-01, Faculty of Engineering Science, Osaka University, Japan, March, 1988.

[97] K. Ito, *Finite-dimensional compensators for infinite-dimensional systems via Galerkin-type approximation*, SIAM J. Control Optim., 28 (1990), pp. 1251–1269.

[98] C. A. Jacobson and C. N. Nett, *Linear state space systems in infinite-dimensional space: the role and characterization of joint stabilizability detectability*, IEEE Trans. Automat. Control, 33 (1988), pp. 541–550.

[99] E. W. Kamen, P. Khargonekar, and A. Tannenbaum, *Stabilization of time-delay systems using finite-dimensional compensators*, IEEE Trans. Automat. Control, 30 (1985), pp. 75–78.

[100] P. P. Khargonekar, I. R. Petersen, and M. A. Rotea, *H^∞-optimal control with state feedback*, IEEE Trans. Automat. Control, 33 (1988), pp. 786–788.

[101] P. P. Khargonekar, H. Ozbay, and A. Tannenbaum, *The four-block problem; stable planes and rational weights*, Internat. J. Control, 50 (1989), pp. 1013–1023.

[102] P. P. Khargonekar and K. Poolla, *Robust stabilization for distributed systems*, Automatica—J. IFAC, 22 (1986), pp. 77–84.

[103] P. P. Khargonekar and E. D. Sontag, *On the relation between matrix fraction factorizations and regulable realizations of linear systems over rings*, IEEE Trans. Automat. Control, 27 (1982), pp. 627–638.

[104] I. Lasiecka and R. Triggiani, *Differential and algebraic Riccati equations with applications to boundary and point control problems: Continuous theory and approximation theory*, Lecture Notes in Control Inform. Sci. 164, Springer-Verlag, Berlin, New York, 1991.

[105] K. Lenz, H. Ozbay, A. Tannenbaum, J. Turi, and B. Morton, *Frequency domain analysis and robust control design for an ideal flexible system*, Automatica—J. IFAC, 27 (1991), pp. 947–962.

[106] H. Logemann, *On the existence of finite-dimensional compensators for retarded and neutral systems*, Internat. J. Control, 43 (1986), pp. 109–121.

[107] ——, *Funktionentheoretische Methoden in der Regelungstheorie Unendlichdimensionaler Systems*, Ph.D. thesis, 1986, Report 156, Institut für Dynamsiche Systeme, Universität Bremen, Bremen, Germany.

[108] ——, *On the transfer matrix of a neutral system: characterizations of exponential stability in input-output terms*, Systems Control Lett., 9 (1987), pp. 393–400.

[109] ——, *New results on the Nyquist criterion and applications to robust stabilizability for infinite-dimensional systems*, in Robust Control of Linear Systems and Nonlinear Control, M. A. Kaashoek, J. H. van Schuppen, and A. C. M. Ran, eds., Proc. Internat. MTNS-89, Vol. II, Birkhäuser, Basel, Switzerland, 1990.

[110] ——, Private communication.

[111] ——, *Circle criteria, small-gain conditions and internal stability for infinite-dimensional systems*, Automatica—J. IFAC, 27 (1991), pp. 677–690.

[112] H. Logemann and D. H. Owens, *Robust high-gain feedback control of infinite-dimensional minimum phase systems*, IMA J. Math. Control Inform., 4 (1987), pp. 195–220.

[113] ——, *Low-gain control of unknown infinite-dimensional systems: a frequency domain approach*, Dynamics Stability Systems, 4 (1989), pp. 13–29.

[114] D. McFarlane and K. Glover, *Robust controller design using normalized coprime factor plane descriptions*, Lecture Notes in Control and Inform. Sci., 138, Springer-Verlag, Berlin, New York, 1990.

[115] D. Meyer and G Franklin, *A connection between normalized coprime factorizations and linear quadratic regulator theory*, IEEE Trans. Automat. Control, 32 (1987), pp. 227–228.

[115a] C. N. Nett, *The fractional representation approach to robust linear feedback design: a self-contained exposition*, MSc. thesis, ESCS Dept., Rensselaer Polytechnic Institute, Troy, NY, 1984.

[116] C. N. Nett, C. A. Jacobson, and M. J. Balas, *Fractional representation theory: robustness with applications to finite-dimensional control of a class of linear distributed systems*, Proc. IEEE Conf. on Decision and Control, 1983, pp. 269–280.

[117] ——, *A connection between state space and doubly coprime fractional representations*, Trans. IEEE, 9 (1984), pp. 831–832.

[118] S. A. Nefedov and F. A. Sholokhovich, *A criterion for the stabilizability of dynamical systems with finite-dimensional input.* Differentsial'nye Uravneniya, 22, Plenum, New York, 1986, pp. 163–166.

[119] H. Nijmeijer and J. M. Schumacher, eds., *Three decades of mathematical system theory*, A Collection of Surveys at the Occasion of the Fiftieth Birthday of Jan C. Willems, Lecture Notes in Control and Inform. Sci. 135, Springer-Verlag, Berlin, New York, 1989.

[120] R. Ober, *A parametrization approach to infinite-dimensional balanced systems and their approximation*, IMA J. Math. Control Inform., 4 (1987), pp. 263–279.

[121] R. Ober and S. Montgomery-Smith, *Bilinear transformation of infinite-dimensional state-space systems and balanced realisation of nonrational transfer functions*, SIAM J. Control Optim., 6 (1990), pp. 438–465.

[122] H. Ozbay, *H^∞-control of distributed systems: An skew Toeplitz approach*, Ph.D. thesis, University of Minnesota, Minneapolis, MN, June, 1989.

[123] ——, *Controller reduction in the two-block H^∞-optimal design for distributed plants*, Internat. J. Control, 54 (1991), pp. 1291–1308.

[123a] ——, *H^∞-optimal controller design for a class of distributed parameter systems*, Automatica, to appear.

[124] H. Ozbay, M. C. Smith, and A. Tannenbaum, *Mixed sensitivity optimization for unstable infinite-dimensional systems*, SIAM J. Control Optim., submitted.

[125] L. Pandolfi and A. W. Olbrot, *On the minimization of sensitivity to additive disturbances for linear distributed parameter* MIMO *feedback systems*, Internat. J. Control, 43 (1986), pp. 389–399 (corrections: 47 (1988), pp. 1161–1164).

[126] J. R. Partington, *Approximation of unstable infinite-dimensional systems using coprime factors*, Systems Control Lett., 16 (1991), pp. 89–96.

[127] J. R. Partington and K. Glover, *Robust stabilization of delay systems by approximation of coprime factors*, Systems Control Lett., 14 (1990), pp. 325–333.

[128] J. R. Partington, K. Glover, H. J. Zwart, and R. F. Curtain, *L_∞-approximation and nuclearity of delay systems*, Systems Control Lett., 10 (1988), pp. 59–65.

[129] PCMATLAB, The Math. Works, Inc., Sherborn, MA, 1989.

[130] A. J. Pritchard and D. Salamon, *The linear quadratic optimal control problem for infinite dimensional systems, Part* I*: A semigroup theoretic approach for systems with unbounded input and output operators*, SIAM J. Control Optim., 25 (1987), pp. 121–144.

[131] ——, *Part* II*: Retarded systems with delays in control and observation*, IMA J. Math. Control Inform., 2 (1985), pp. 335–362.

[132] A. J. Pritchard and S. Townley, *A Stability Radius for Infinite-Dimensional Systems*, Distributed Parameter Systems, Proc. 3rd Internat. Conf. Vorau, Austria, Lecture Notes in Control and Inform. Sci. 102, Springer-Verlag, Berlin, 1987, pp. 272–291.

[133] A. J. Pritchard and S. Townley, *Robustness of linear systems*, J. Differential Equations, 77 (1989), pp. 254–286.

[134] ——, *Robustness optimization for uncertain infinite-dimensional systems with unbounded inputs*, IMA J. Math. Control Inform., to appear.

[135] A. A. Rodriguez and M. A. Dahleh, *Weighted H^∞-optimization for stable, infinite-dimensional systems using finite-dimensional techniques*, Proc. 29th Conf. on Decision and Control, IEEE, New York, 1990, pp. 1814–1820.

[136] D. Salamon, *Control and Observation of Neutral Systems*, Res. Notes in Math. 91, Pitman, Boston, MA, 1984.

[137] ——, *Infinite-dimensional linear systems with unbounded control and observation: a functional analytic approach*, Trans. Amer. Math. Soc., 300 (1987), pp. 383–431.

[138] ——, *Realization theory in Hilbert space*, Math. Systems Theory, 21 (1989), pp. 147–164.

[139] M. C. Smith, *On stabilization and the existence of coprime factorizations*, IEEE Trans. Automat. Control, 34 (1989), pp. 1005–1007.

[140] ——, *Well-posedness of H^∞-optimal control problems*, SIAM J. Control Optim., 28 (1990), pp. 342–358.

[141] G. Tadmor, *Worst-case design in the time domain. The maximum principle and the standard H^∞-problem*, Math. Control Signals Systems, 3 (1990), pp. 301–324.

[142] ——, *The standard H-infinity problem and the maximum principle: the general linear case*, manuscript.

[143] S. Townley, *Robustness of infinite-dimensional systems*, Ph.D. thesis, University of Warwick, Warwick, U.K., 1987.

[144] B. van Keulen, M. Peters, and R. F. Curtain, *H^∞-control with state feedback: the infinite-dimensional case*, J. Math. Systems, Estimation and Control, to appear.

[145] B. van Keulen, *H^∞-control with measurement feedback for linear infinite-dimensional systems*, J. Math. Systems, Estimation and Control, to appear.

[146] M. Vidyasagar, *Control System Synthesis: A Factorization Approach*, MIT Press, Cambridge, MA, 1985.

[147] M. Vidyasagar and H. Kimura, *Robust controllers for uncertain linear multivariable systems*, Automatica—J. IFAC, 22 (1986), pp. 85–94.

[148] M. Vidyasagar, H. Schneider, and B. A. Francis, *Algebraic and topological aspects of feedback stabilization*, IEEE Trans. Automat. Control, 27 (1982), pp. 880–894.

[149] S. de Vries, *A frequency domain analysis of models of flexible beams*, Masters' Project, 1987, Mathematics Institute, University of Groningen, Groningen, the Netherlands.

[150] G. Weiss, *Admissibility of unbounded control operators*, SIAM J. Control Optim., 27 (1989), pp. 527–545.

[151] ——, *The representation of regular linear systems on Hilbert spaces*, in Distributed Parameter System, Proc. of the Conference in Vorau, Austria, July 1988, F. Kappel, K. Kunisch, and W. Schappacher, eds., Birkhäuser, Basel, 1989, pp. 401–416.

[152] ——, *Admissible observation operators for linear semigroups*, Israel J. Math., 65 (1989), pp. 17–43.

[153] ——, *Admissibility of input elements for diagonal semigroups on l^2*, Systems Control Lett., 10 (1988), pp. 79–82.

[154] J. C. Willems, *The Analysis of Feedback Systems*, MIT Press, Cambridge, MA, 1971.

[155] Y. Yamamoto, *Pseudo-rational input/output maps and their realizations: a fractional representation of infinite-dimensional systems*, SIAM J. Control Optim., 26 (1988), pp. 1415–1430.

[156] Y. Yamamoto, *Equivalence of internal and external stability for a class of distributed systems*, Math. Control Signals Systems, 4 (1991), pp. 391–410.

[157] Y. Yamamoto and S. Hara, *Relationships between internal and external stability for infinite-dimensional systems with applications to a Servo problem*, Proc. 28th IEEE Conference on Decision and Control, Los Angeles, CA, 1987.

[158] G. Zames, *Feedback and optimal sensitivity: model reference transformations, multiplicative seminorms and approximate inverses*, IEEE Trans. Automat. Control, 26 (1981), pp. 301–320.

[159] Z. Zhou and P. P. Khargonekar, *An algebraic Riccati equation approach to H^∞-optimal control*, Systems Control Lett., 11 (1988), pp. 85–91.

[160] ——, *On the weighted sensitivity minimization problem for delay systems*, Systems Control Lett., 8 (1987), pp. 307–312.

[161] S. Q. Zhu, *A graph metric for a class of* MIMO *linear distributed systems, analysis and optimization of systems*, INRIA, A. Bensoussan and J. L. Lions, eds., Lecture Notes on Control and Inform. Sci. 111, Springer-Verlag, Berlin, 1988.

[162] ——, *Graph topology and gap topology for unstable plants*, IEEE Trans. Automat. Control, 34 (1989), pp. 848–855.

[163] ——, *On normalized coprime fractions of distributed* LTI *systems*, IEEE Trans. Automat. Control, 36 (1991), pp. 489–491.

[164] ——, *Robustness of Feedback Stabilization: A Topological Approach*, Ph.D. thesis, Eindhoven, the Netherlands, 1989.

[165] S. Q. Zhu, M. L. J. Hautus, and C. Praagman, *Sufficient conditions for robust* BIBO *stabilization: given by the gap metric*, Systems Control Lett., 11 (1988), pp. 53–60.

[166] H. J. Zwart, R. F. Curtain, J. R. Partington, and K. Glover, *Partial fraction expansions for delay systems*, Systems Control Lett., 10 (1988), pp. 235–244.

Index